SECOND EDITION

REVOYAGE OF THE MAYFLOWER

Societal Values – Conservation's Driving Force

Herbert A. Raffaele

OTHER BOOKS BY HERBERT A. RAFFAELE

Birds, Beasts, & Bureaucrats

Birds of the West Indies

A Guide to the Birds of the West Indies

Wildlife of the Caribbean

A Guide to the Birds of Puerto Rico and the Virgin Islands

TABLE OF CONTENTS

FOREWORD

Half a century has elapsed since the first Earth Day and the birth of the modern environmental movement in 1970. The decade that followed established powerful environmental legislation. Laws to purify our air, clean our water, save our endangered species, and sustain our environment, were promulgated nationwide. A green generation was born, and the political establishment took notice. The United States seemed on the cusp of a bright, prosperous, and environmentally sensitive future.

Now, 50 years later, our nation and our planet seem ravaged on every front. Concern for pure air and clean water has scant political traction. Species that the Endangered Species Act sought to restore, such as the gray wolf and grizzly bear, are hunted remorselessly at every opportunity. No strong environmental legislation has been passed for decades. Moreover, climate change threatens to torment us with unfathomable environmental calamities of a magnitude inconceivable half a century ago – calamities brought on by our own actions founded on greed and lack of caring for the planet.

So, how did we get here? What went wrong?

Prior to the first Earth Day, the fledgling environmental movement was fragmented. Groups that had been fighting individually against oil spills, polluting factories and power plants, raw sewage, toxic dumps, pesticides, freeways, the loss of wilderness, the extinction of wildlife and more united on Earth Day around their shared values. To this day, collaboration and teamwork are at the core of the movement.

Unfortunately, some of the core messaging from the first Earth Day – such as the call by Governor Forrest Anderson of Montana for a cultural revolution – has fallen by the wayside. In its place we find efforts to preserve the landscape or save species by any of many means – legislation, enforcement, purchase, or whatever – without adequate attention to the underpinnings that make such efforts necessary, our society's cultural values as they relate to conservation. As pointed out so eloquently by Leopold long ago, and suggested by others since, here we have the crux of the matter – the need for a conservation ethic. Such an ethic has been all but ignored by society to the point where actively saving our environment is not ingrained into day-to-day life. Somewhere along the line, humans formed the wrong idea about our place in the world. We will not see the error of our ways until we reinvent how we live our lives.

But here's what I feel good about. I'm excited to see an innovative, impactful case made for getting resource conservation on track. To see that the science necessary to achieve conservation focuses not just on animals and the environment but revolves around people and society. To see a precise framework laid out, a roadmap if you will, for creating the culture change so badly needed for conservation to be effective. To see profound thinking, not just hand-waving, that builds upon Leopold's musings regarding a conservation ethic, and leads us on the road to creating it. To see the power of *not* telling us what our conservation ethic should be, but rather helping us recognize that it is something *we* must build – together – from the ground up, not the top down, community by community. To see that there is a real opportunity to change the world for the better if we would only unlock our minds to consider how we – the conservation community, civic organization, or the caring citizen – must conduct our business and our lives in a more strategic and impactful way.

For these reasons, I am both proud and honored to write this forward to *Revoyage of the Mayflower: Societal Values – Conservation's Driving Force*. What a visceral message – the parable of sailing the *Mayflower* with emigrants from India aboard rather than Pilgrims. Who would have thought that North America's fauna would be so different today but for a modified ship's manifest reflecting the cultural origins of the 102 passengers who suffered that challenging voyage. The result would have been so much more positive! *Revoyage of the Mayflower* is full of insights and replete with fascinating stories based on the author's personal experiences all over the world with iconic wildlife. Most importantly, it provides us a precise blueprint for a better tomorrow.

As President of EARTHDAY.ORG, I call upon all our partners, others interested in conservation, and everyone who cares about making the earth a better place, to take this book to heart. Following its guidance empowers both individuals and groups to truly make a difference. I believe that the strategies laid out in *Revoyage of the Mayflower* – identifying and framing conservation values, customizing efforts to reach each sub-sector of society, and engaging all community members, among others – will help launch the cultural change long-overdue yet essential for conservation efforts to be truly effective. Ultimately, my great hope is that it will also lead to Earth Day being celebrated ***not*** once a year, but truly *every* day.

Kathleen Rogers
President, EARTHDAY.ORG

PREFACE

Why another conservation book? There are so many of them. And many are so much more eloquent than this one. But I have not written this to be eloquent. I've written this book to provide solutions. Simply, I have become weary of reading beautiful prose that elucidates the problems facing the environment around the world but lacks a clear vision of how to address them. Many call for "hope," a term which can be uplifting when things are looking bad – as they are at present – but which is hardly a solution. "Hoping" that it does not rain scarcely makes a difference as to whether it will or not! Consequently, I am not a believer in hope – and it is not among the solutions I propose here.

Why me? I have spent my entire career involved in conservation, beginning as a field biologist having the time of my life on the ecologically wonderful but politically complicated island of Puerto Rico. After seven years there and another cluster procuring a PhD in ecology, I had the good fortune to spend 12 years coordinating conservation programs in Latin America and the Caribbean for the US Fish and Wildlife Service. Subsequently, I spent 18 years managing the Service's international conservation efforts, absent the trade component. What did this entail? It meant providing annual grants to conserve endangered flagship species such as tigers, elephants, gorillas, and sea turtles – that includes revenues generated from the endangered species stamp (with the face of a tiger on it) many of you will find available in your local post offices. It meant implementing bilateral treaties and associated conservation programs with Mexico, Russia, Canada, and Japan. It meant leading US engagement in multilateral treaties such as the Convention on Wetlands of International Importance and the Convention on Nature Protection and Wildlife Preservation in the Western Hemisphere among others. It meant managing a staff of close to 30 individuals and an annual budget that maxed out at over $22 million per year – most of which was spent on conservation projects all over the globe. During the course of my career, our programs funded many hundreds of proposals and reviewed many thousands. Some of these were with BINGOS (Big International Non-Governmental Organizations) while others supported local groups. These experiences provided me a unique perspective on what works and what does not on the ground. Based on this background, I lean heavily on wildlife conservation in the examples I present, particularly taken from abroad – a reflection of where my core experience lies. The book and its proposed solutions, however, are germane to all aspects of environmental conservation.

Who is this for? If you are concerned about what you see happening in the world around you – to its wildlife – to nature in general, and – to the global environment – then I think you will find this book worthwhile. My intention is to offer a different perspective along with concrete solutions to our country's and the world's conservation dilemma. Instead of bemoaning the sad state of environmental affairs, or being alarmist about the need for action, this book aims to provide solutions. This book should serve any individual, community, or other entity with even a passing interest in environmental conservation. In particular, it should benefit numerous conservation organizations with their hearts in the right place, but short on understanding what drives conservation and what it takes to really make a difference. Those that shift to addressing underlying causes of the problem rather than conspicuous symptoms will have that much more of a positive impact. It should also serve enlightened communities – communities large or small that care about the environment in which they live. It is these entities that have the flexibility, positive aspirations, and shared identity necessary to use the tools presented here to advance their environmental fortunes in a truly positive way.

Also, though written from a US perspective, the audience for solutions proposed in this book is global. I cannot emphasize this enough. The underlying problems of environmental conservation are the same everywhere; it is their cultural context that differs from place to place. It is very possible that the conservation tools suggested here will be embraced abroad before they are accepted in the United States. That would not be surprising. What is important is that they be used, that their use will confirm their value, and that over time necessary changes in conservation practice will occur more widely.

If anyone should read this book, it is the staffs of the 50 state fish and game agencies in the United States. Those entities, a huge and powerful swath of the US conservation community, are steeped in culture, the central theme of this book. But they function in an undemocratic way. The aim of many is to sustain hunting and angling at all cost despite a changing national demographic increasingly drawn to other outdoor interests. This unwillingness to change has made state agencies increasingly irrelevant to the general public, placed them at cross-purposes to many state citizenries, and generally set back the evolution of conservation practice in the US to an extraordinary extent. To be fair, they are not alone, they just stand out because of their inordinate size, authority, and resources.

A note: This book was originally intended simply as my reflections on achieving conservation based upon my experience. Such a work is counter to most contemporary books that are packed with citations. It was specifically over this issue that a distinguished publisher and I parted ways so that I could write as I pleased. Though I have, in the end, included a number of citations, they are not comprehensive, as my intent was never to write an academic work. To the extent that academicians revile this approach, I can only say that they and I do not possess the same reality. Despite all the modern journals on conservation values, conservation ethics, and the like, academicians seem to live in an alternate universe that rarely crosses paths with major aspects of conservation practice on the ground – an all-too-common, but very regrettable circumstance. That said, the sparsity of citations is in no way intended to suggest that all the ideas in this book are novel and have never been touched upon by others. What I do believe is novel, however, is the compilation in this book of a comprehensive conceptual framework that provides powerful tools for more effective conservation action.

PREFACE TO SECOND EDITION

During a recent trip to India in May 2023 I had the opportunity to spend two weeks at the Ashoka Trust for Research in Ecology and the Environment (ATREE) in Bangalore thanks to the generosity of Dr. Kamal Bawa, its president, who recognized the importance of Revoyage of the Mayflower's message for conservation students and professionals within that country. The experience was enlightening in several ways. One was when an Indian graduate student responded to a question regarding how many bird species are extinct in India given that there are 9 extinct birds in the United States. Her response was, "There must be hundreds." In actuality, there are only two. Why her response? I believe it is an artifact of hundreds of years of colonialism. But, perhaps of more importance, the student's response went to show how out of touch Indians themselves are with their own extraordinary wildlife heritage.

A second experience concerned references to Hinduism that I made in the book. These references were interpreted as short-changing the contributions to conservation of various other ethnic groups in India, particularly tribal peoples who live in more remote localities and consequently make the greatest sacrifices in coping with wild animals that affect their lives and livelihoods. The first edition made no reference to such groups so essential to the survival of India's wildlife. This oversight demanded immediate attention and so this new edition.

To take full advantage of this rewrite I have tinkered with other elements of the book, most significantly adding a discussion of poisonous snakes and restructuring the section on extinct birds.

INTRODUCTION
RACING DOWN THE WRONG PATH

Recently, I was informed that a former associate of mine, an expert field biologist, now works for a billionaire to implement conservation projects in Africa. What could be more wonderful than that? The job of a lifetime! All the money you could want, scant bureaucracy, freedom to focus on factors that really make a difference to benefit conservation. These efforts, I was told, address the acquisition of prime habitat for elephants and other wildlife, animals severely persecuted in increasing portions of their ranges. Purchased lands will then be patrolled to deter poaching, the entire effort undertaken in conjunction with local authorities so that hopefully, over time, these reserves can be transferred to them for protection in perpetuity. Amazing! Conservation at its best!! More projects like this and the world's endangered species can be saved.

Or can they?

I regret to say, I think not. Looks can be terribly deceiving.

In fact, given what is increasingly occurring regarding wildlife conservation in the US and around the world over the past century, conservation efforts such as this, as popular, seemingly well-focused, measurable, and problem-oriented as they are, simply do not work – and that was prior to the added complexities of climate change which throws a major monkey wrench into the practice of land management. Moreover, based upon what we now know, though all too often refuse to see, I believe such initiatives are no less than anti-social escapist behavior – a denial of the real world out there that is teetering off-kilter more precariously every day. Efforts such as this are invariably out of sync with the values of local peoples who occupy adjacent lands – not to mention the political seats of power. They are a seed of local resentment which simmers beneath the surface due to the vast sums of money and high-rolling players involved. They are, as I see them, basically a beneficent form of neo-colonialism. Thus, all too regrettably, these initiatives are destined to failure.

Sounds extreme?

I too would have thought so not many years back. And why not. The conservation community has many myths associated with it, embraced by the general populace and professionals alike.

This book aims to dissolve some myths. Not all of them, but certainly some important ones. And in doing so, I hope to make it apparent that rather than extreme, my assertion is soundly based. More important is that present myths are replaced with solid principles for achieving effective, long-lasting conservation. I synthesize these in later chapters into nine fundamental elements that form a framework for success.

Myths die hard. So, it will take a while to make the case. To that end, we shall take a glimpse at the culture of India, set the *Mayflower* to sail again, compare wildlife conservation in the United States and India, and discuss social marketing among other themes. But, in the end, things will all come together in the form of powerful conservation solutions.

While grappling with this manuscript, I have enjoyed the good fortune of, from time to time, gazing out over the great richness of Cape Cod, a remarkable peninsula, shaped like a flexed arm, that stretches far into the Atlantic Ocean off the Massachusetts coast. The Cape's history, both culturally – from the arrival of the Pilgrims, to its days as a whaling hub, to its present state as a vacation destination – and naturally – as the epicenter of Northern right whale breeding and cod fishing – is a fascinating saga. As it so happens, the preeminent event in the Cape's cultural history – the arrival of the Pilgrims – will serve in this book as a prime example to illustrate the fundamental importance of societal values in the conservation of the world's living resources. This is a factor so important that few others matter.

We shall get to that in due course, but first let's consider one of the Cape's less noted, but still special attributes, one of many that led to one-third of the Cape being set aside as a national seashore under the jurisdiction of the US National Park Service. It happens to be a controversial, little bird.

Is "Poisoning" the Best Way to Save the Piping Plover?

Along the sandy outer beaches of the Cape breeds a shorebird, known as the piping plover. This tiny, white puffball of feathers with orange legs and bill, and various black markings, while on its breeding grounds in Massachusetts and other East Coast states, is entirely dependent upon sandy ocean beaches for its existence. Tame in demeanor, the bird is not difficult to find as it skitters among the detritus and beach grasses above the high tide line. With luck, even its nest, nothing more than an indentation in the sand with a few pebbles and seashells lining it, and perhaps containing a few incredibly camouflaged eggs or young, can also be found.

The piping plover is not alone in its desire for beachfront real estate. Such beaches are a prime recreational and living environment for our own species which, as a result of ceaseless expansion into this habitat, along with careless management of the same, has ultimately led to this plover's decline – one that landed the bird on the US Endangered Species List in 1986. Being on the Endangered Species List has a number of consequences, one of which is that land management agencies, such as the US National Park Service, are responsible for making efforts to improve the status of listed species. Regrettably, efforts by the National Park Service had, in the past, proven less than successful. Resultantly, but a few years back, Cape Cod National Seashore

(CCNS) managers proposed a dramatic new plan for restoring plover numbers. In a nutshell, the plan went something like this. Nest success of piping plovers was failing due to predators such as coyotes, crows, and foxes. These predators were occurring in unnatural abundance in plover nesting areas because they were "subsidized predators." (Subsidized predators are those which benefit from human food wastes left behind and are able to survive and reproduce in increased numbers.) An informational campaign to stop visitors from being careless with wastes had failed and so the new alternative was that these predators be killed, mostly by poisoning. The Cape Cod National Seashore was high on this solution. In fact, the entity was so positive the plan would succeed that in synchrony with the poisoning CCNS would also allow more vehicles on the beach. Without a doubt, these vehicles would penetrate to areas not conveniently reached on foot. Yet, somehow the CCNS was unconcerned that additional vehicles would exacerbate the trash problems – thus exacerbating the subsidized predator problem. Perhaps this was because CCNS could just kill off any new nuisance animals. (Or, perhaps it was more concerned about the revenues from the permits allowing vehicular beach access.) The Cape Cod National Seashore was also unbothered by this proposal discounting the primary interests of the four million visitors to the seashore each year who found the natural environment there an important attraction, were willing to sacrifice for its benefit, and with all likelihood would have found the predator poisoning proposal abhorrent. These survey results were from the CCNS's own findings.

I mention this predator-poisoning solution to the piping plover dilemma because it is just such mishandling of conservation problems that this book aims to address. Generally, in the US in particular, natural resource managers, whether federal, state, or within many private organizations, have an inadequate understanding of how best to conserve the wildlife and natural resources they are trying to protect. To be clear, this failure is not a matter of not caring. The vast majority of natural resource managers would not have entered this field absent a concern for wildlife and nature. The failure hinges on our culture and the history of conservation in this country – a lengthy legacy of missteps and misunderstanding of what conservation is all about. This book attempts to provide a more holistic and pragmatic perspective.

PIPING PLOVER

Modern Society's Disconnectedness from the Natural World

Human beings and wildlife have coexisted since the origin of our species. During the dawn of that relationship, when the human footprint, both literally and figuratively, was scant indeed, the need for a concept such as "conservation" was irrelevant. Simply, survival was the order of the day. It is now widely believed that well before the advent of agriculture, while *Homo sapiens* were solely hunters and gatherers, our species managed to affect the extinction of numerous species ranging from some of the wondrous mammals of the Pleistocene such as the wooly mammoth, mastodon, and giant ground sloth, to what were then the largest birds in the world, the flightless elephant bird of Madagascar, nearly three times the weight of the modern day ostrich, and the moas of New

Zealand, the tallest of which stood a full 12 feet.

Though the hunters and gatherers of thousands of years ago appear to be responsible for the extinction of some spectacular species, for them to survive over time, it was essential that they develop conservation values that respected the earth and the creatures upon it. This is because, while the first peoples to colonize a new land had the luxury of untapped fauna and flora on which to grow and prosper, overexploitation of these resources would have led subsequent generations to adjust to the declining availability of wildlife and plants on which they depended. Those peoples who recognized this would have been forced to develop a set of mores which enabled them to live in a sustainable relationship, an equilibrium so to speak, with the living things and the land around them. Those peoples who did not recognize this, or who chose to ignore the shortcomings of their unsustainable relationship with their basic sources of food and fiber, were destined to either die of famine or move on to a new locality. The development of a powerful ethic connecting early human cultures to the nature around them was essential if any of those societies were to endure. As George Tinker (2010), professor of American Indian Cultures and Religious Traditions at the Iliff School of Theology explains, "An Indian environmental concern begins with a deeply embedded sensitivity to our relationships with all life-forms, meaning all persons – if we can, as Indians do, interpret the English word 'person' much more broadly, to include other-than-human persons." This means that other living things were treated as kin, thus with appropriate respect.

To many observers, modern civilizations have managed to endure while their perspective on life has strayed further and further from connecting to the land and the numerous resources associated with it. The development of agriculture, animal husbandry, and silviculture have enabled us to manage our sources of food, fiber, drugs, and most other human needs to the point where the extinction of most wild species of plants and animals has become, from a narrow livelihood perspective, nearly irrelevant. What better example might we find in the United States than the extinction of the passenger pigeon? Once perhaps the most abundant bird in the world, with population estimates ranging from 3-5 billion, the species was hunted commercially with such intensity that train cars were packed full with these small birds destined for the voracious food markets of America's growing cities. Yet the disappearance of this formerly important food source caused scarcely a blip to register on the growth and prosperity of the nation.

You might ask how many of our populace today remember this bird. How many miss it? Many towns and villages east of the Mississippi have names associated with this pretty dove – Pigeon Creek in Monongahela, Pennsylvania, the town of Pigeon Forge in Tennessee, the Pigeon River in Wisconsin, to name a few. Do residents know the derivation of these monikers? Unquestionably some do, but to most the passenger pigeon is virtually a lost memory – this despite the bird having provided much-needed sustenance in the earliest days of European colonization of this continent and even to Revolutionary War patriots during the fight for independence.

This is to say that sustainable conservation of living resources, formerly a necessity for hunter-gatherer societies to survive over time, has, in our modern era, played an ever-diminishing role. Certainly many species, habitats, and ecosystems serve some important purpose or provide a particular service to humankind. Despite this circumstance, in modern industrial societies, we're not dependent on whether there is a nearby woodlot for fuel, or game for dinner, or fish in a neighboring stream.

The trend is clear. As modern society becomes more technologically advanced, dependence on wilderness and the living things in our immediate surroundings inevitably becomes less important for our survival. In the present day, the natural world around us already verges on being an afterthought to many people. To many, sustaining the natural world is likely perceived more as a luxury for the enlightened and a playground for the well-to-do than as an underlying necessity for the long-term viability and health of society. Outdoor philosopher Kate Rawles (2010) observes, a "critical flaw in the industrialized world view is the portrayal of human beings as somehow separate from the rest of the natural world."

The implications of this circumstance are vast. Increasingly, due to improved technologies and invention, our awareness of, as well as direct dependence on, the natural world is diminishing. Increasingly, population growth and rising demand for consumer goods place greater and greater stress on natural environments. In fact, in the entire history of humankind, it took until the 1950s to build the world economy to a level of $7 trillion. Presently, the world economy grows by that amount in a mere decade (Speth 2010). Trends towards urbanization are separating more and more people from daily contact with the rhythms of nature, while the attraction of television and the allure of computer games have made the present generation of youth the most indoor-oriented in the history of humankind. Stated in other words by Stephen Kellert (2010), professor of social ecology at the Yale School of Forestry and Environmental Studies, "Modern humans have in many ways lost their bearings as biological beings, as just another animal and species in the firmament of creation... we are even more separated, if not alienated, from our biological roots." Phrased lyrically by the environmental writer and poet Wendell Berry (2010), "As industrial technology advances... it carries people away from where they belong... and it destroys the landmarks by which they might return." So severe has this problem become that the term "nature deficit disorder" has been coined to reflect its impact on today's children. For a comprehensive analysis of this issue, see Richard Louv's 2009 book *Last Child in the Woods*.

These developments do not bode well for our future. The challenges faced by those of us who realize that modern society must find a means to recognize the value of the natural world and learn to live sustainably within it are greater and more complex than ever before. Failure to achieve this goal forebodes the loss of ever-increasing numbers of species, runaway climate change, expanding deserts, and depletion of natural resources. Such prospects increasingly threaten our existence year by year, day by day.

As each future generation is more noticeably removed from nature and dependence upon it for survival, it can be expected that such losses will be perceived with increased apathy. This decimation of nature is a tragedy of potentially epic proportions and one which neither we, nor any future generations to come, may have the capacity to reverse.

Our Conservation Community Is Not Up to the Challenge

Highly regrettable is that many conservation efforts, both at home and abroad, are not up to the challenge we have created for ourselves.

"However beautiful the strategy, you should occasionally look at the results." – WINSTON CHURCHILL

Let me offer a few examples to give you a sense of the problems facing a handful of taxa and natural systems:

- **Birds:** According to a recent analysis, bird numbers in North America have declined by nearly 30 percent in the last 50 years.

New research published in the journal *Science* shows massive losses among US bird populations – with steep declines in every habitat (Rosenberg et al. 2019).

• **Bats:** "Current estimates of bat population declines in the northeastern US since the emergence of White-nose Syndrome (WNS) are approximately 80 percent. This sudden and widespread mortality associated with WNS is unprecedented in hibernating bats, among which disease outbreaks have not been previously documented. It is unlikely that species of bats affected by WNS will recover quickly because most are long-lived and have only a single pup per year. In temperate regions, bats are primary consumers of insects, and a recent economic analysis indicated that insect suppression services (ecosystem services) provided by bats to US agriculture is valued between $4 to 50 billion per year. Despite efforts to contain it, WNS continues to spread" (National Wildlife Health Center, USGS website).

• **Amphibians:** "Amphibian declines are a global phenomenon that has continued unabated in the United States since at least the 1960s. Declines are occurring even in protected national parks and refuges. The average decline in overall amphibian populations is 3.79 percent per year, though the decline rate is more severe in some regions of the US, such as the West Coast and the Rocky Mountains. If this rate remains unchanged, some species will disappear from half of the habitats they occupy in about 20 years" (ibid.).

• **Trees:** Forests are under attack from invasive species, diseases, and unprecedented outbreak of pests, while trying to withstand stress caused by climate change and drought. The Midwest is fighting the invasive species emerald ash borer, which is killing tens of millions of ash trees. New England has seen tens of thousands of trees succumb to the Asian long-horned beetle, which, if it spreads, is estimated to be able to destroy 30 percent of the country's hardwoods. In the West, millions of trees are being lost to the combined threat of mountain pine beetles and white pine blister rust. Cities across the country have lost tens of thousands of elm trees to Dutch elm disease over the last 60 years (American Forester website).

- **Bees:** "Data from the 2012-2013 winter, indicate an average loss of 45.1 percent of hives across all US beekeepers, up 78.2 percent from the previous winter, and a total loss of 31.1 percent of commercial hives, on par with the last six years. In total, bees contribute more than $15 billion to US crop production" (Holland 2013).
- **Natural systems:** Many of these are failing or degrading at an alarming rate. Coral reefs, other ocean ecosystems, rain forests, biological diversity, global temperatures, flooding, major storms, climate change generally, and many more. Numerous books and reports document this tragic situation, so we will not delve into it.

"What we do in the world flows from how we interpret the world." - CHARLES BIRCH

The grave situation summarized above is commented on bluntly by renowned environmental author Paul Hawken (2010): "Every living system is declining, and the rate of decline is accelerating. Kind of a mind-boggling situation." It has gotten so bad that Geist (1995) has proposed a "Noah's ark" approach wherein biota from the earth's highest latitudes and altitudes, those he believes most susceptible to on-going climatic changes including ultraviolet radiation, would be housed in large, roofed structures to avoid them becoming extinct. Referring to the 21st century, Holmes

Rolston (1994) suggests "the worry... is that humans may destroy their planet and themselves with it."

> *"Humanity is conducting an unintended, uncontrolled, globally pervasive experiment whose ultimate consequences could be second only to a global nuclear war."*
>
> – WORLD METEOROLOGICAL ORGANIZATION, UNEP, AND ENVIRONMENT CANADA

Why is the Conservation Movement Failing?

Why is the conservation movement as a whole failing? Overall, the central shortcoming of most present-day conservation efforts is their lack of adequate attention to what matters most in achieving conservation – our values system, the single most important factor in determining our actions as individuals and as a society. As ecologist and writer David Orr (1992) states so firmly, "mainstream scholars who trouble themselves to think about disappearing species and shattered environments appear to believe that cold rationality, fearless objectivity, and a bit of technology will get the job done. If that were the whole of it, however, the job would be much further along than it is now." In his parting remarks as editor of the journal *Conservation Biology*, David Ehrenfeld (1993) highlighted this point succinctly when he stated that "science alone does not have and never will have solutions to the fundamental environmental problems of our time, which are religious in the largest sense of the word, dealing as they do with values and the human spirit. If we remember this at all times, our science will then be free to play the part that is expected of it in the battle to save the life on this planet." All too little has changed since the days these observations were written.

Woefully, the conservation community continues its failed over-reliance on studying the animals and habitats it wants to conserve and the dissemination of information based upon the misconception that providing the correct facts to an individual will result in the making of wise decisions (see Heberlein 2012). This is not to say that wildlife science and information are not important; of course they are. But we need more social sciences, like conservation psychology involved, because the need for conservation is a human-caused problem (Cooney 2011). We have the wrong balance. The inordinate attention to wildlife rather than societal research is misplaced in most present-day conservation practice. Renowned American conservationist Aldo Leopold pinpointed this long ago, but in doing so he also recognized the great challenge to be surmounted in accomplishing this goal. "One of the anomalies of modern ecology is that it is the creation of two groups each of which seems barely aware of the existence of the other. The one studies the human community almost as if it were a separate entity, and calls its findings sociology, economics, and history. The other studies the plant and animal community, and comfortably relegates the hodge-podge of politics to "the liberal arts." The inevitable fusion of these two lines of thought will, perhaps, constitute the outstanding advance of the present century" (Leopold, unpublished notes).

"The greatest danger in times of turbulence is not the turbulence – it is to act with yesterday's logic."

– PETER DRUCKER

Clearly Leopold recognized the tremendous importance of better engagement between these two fields, but the "inevitable fusion" that he foresaw has yet to happen. As a result, "the outstanding advance" of the 20th century, long overdue, must now await our achievement of it in the 21st.

A telling example of how serious this problem remains can be seen from a 2016 special *"Issues in Ecology"* published by the Ecological Society of America (Evans et al. 2016) offering recommendations on how to update the powerful, but increasingly out of date Endangered Species Act of 1973 (ESA). The article was authored by a cadre of 18 renowned scientists and conservation professionals representing a who's who of prestigious entities ranging from the American Association for the Advancement of Science and auspicious conservation organizations to institutions, such as the National Socio-Environmental Synthesis Center at the University of Maryland. The authors specifically aimed to "identify key successes and shortcomings of recovery programs, and discuss... six broad strategies to increase the effectiveness of ESA implementation."

Recommended Strategies to Improve Effectiveness of the Endangered Species Act (Evans et al. 2016):

• Better prioritizing of species recovery funding;

• Strengthening species recovery partnerships and collaboration among agencies plus creating incentives for private landowners;

• Promoting more monitoring and adaptive management relating to ecological complexity and uncertainty;

• Developing more measurable recovery criteria based on the best available science;

• Using climate-smart conservations strategies; and

• Evaluating ecosystem-based approaches to increase efficiency of managing for recovery.

Do you notice what's missing in the above list? There is no reference whatsoever to addressing the shifting societal values prevalent in the US that have jeopardized the Endangered Species Act for decades and are an ever-increasing threat to its survival. The ESA affects many people, businesses, and communities in various ways. The second recommendation makes a gesture in this direction by suggesting incentives for private landowners, but that hardly reflects the magnitude of the issues this act creates. As Stephen Kellert (1996) of the Yale University School of Forestry and Environmental Studies suggests, "The development of a compelling rationale and an effective strategy for protecting endangered species will require an increasing recognition that most contemporary extinction problems are largely the result of socio-economic and political forces." He adds, "To suggest that the causes of a problem are inextricably woven into the fabric of human society implicitly assumes the need to assign fundamental social and perceptual forces a central role in devising solutions."

Such a misfocus, as we've seen in the case of the ESA, is all too typical of US conservation efforts to this day, and results in a dramatic lack of attention to our individual and societal values, ethics, mores, and norms – at great expense to the outcome of our efforts. In the words of authors Moore and Nelson (2010), "Clearly, information is not enough. What is missing is the moral imperative.... Western society is very good at facts. We aren't as good at values." As the distinguished environmental philosopher Bryan Norton (1987) noted, "If society is to address the entire range of concerns regarding biological diversity, it must reexamine its entire style of life and the values that drive it." Meanwhile, Lynn White (1967) states in his classic paper, "More science and more technology are not going to get us out of the present ecologic crisis until we find a new religion, or rethink our old one."

Unquestionably, other factors matter. One which I find particularly perverse, because of its immense counterproductiveness, is the shortage of democracy in the conservation movement – the disenfranchisement of much of the public. This is particularly the case at the state level where conservation has long been managed for hunting and which has been reticent to adjust to a changing world. We shall touch on this as we look at conservation solutions.

In using the term **"values"** I refer broadly to **a person's or a society's overarching belief system which includes related concepts of attitudes, ethics, and mores** about which social psychologists can define and debate their meaning. It is important to note that while many experts believe that our values change slowly, it is also recognized that they are shaped primarily through learning during one's youth (Manfredo; Teel and Zinn 2009) and subject to change through new experiences (Heberlein 2012).

Concomitant with the over-emphasis on wildlife science and information dissemination is what appears to be a precipitous decline in caring for nature currently underway in the United States as touched upon above. Since shortly before the beginning of the present century, the concern of Americans to leave the earth in good shape for future generations appears, according to a number of polls, to have declined by more than half – from 74 percent in 1996 to 34 percent in 2012 (Nelson 2002; Steinberg 2012). Of equal concern, over the course of these 17 years, this negative trend has been continuous and unabating. Despite this trend, the professional conservation community within my sphere, including federal, state, and a good number of larger non-governmental organizations, appears, in most cases, to be unaware of this growing threat or, if hazily aware of it, incapable of getting their collective heads around the issue and devising a path forward to address it, no less reverse this trend.

I have had the good fortune to spend nearly 40 years of my professional life engaged in wildlife conservation efforts around the globe. This experience has ranged from being challenged to a duel while trying to protect sea turtles in Puerto Rico, to providing project funds for indigenous peoples who signed their grant proposal with thumb-prints, to serving as the head of US delegations to global conferences for the implementation of international treaties. This breadth of experience has enabled me to become familiar with diverse conservation approaches – successes and failures – across numerous lands by both the governmental and private sectors. In turn, this perspective allows me to look at conservation in the US from the outside in. Such a view has gradually fostered my very different perspective on the effectiveness of conservation as practiced in this country and how it ought to be modified. Our discussion in this book focuses heavily on the US primarily because this country serves so significantly as a model for the rest of the world. Thus, once the US gets on track, other countries in the world will follow. At the same time, the conservation practices put forward in this book are applicable anywhere. Some countries, as we shall see, are already far ahead of the US on this front. Entities should assess the recommendations made herein from their own local context and implement them at their own pace.

How We Get There From Here

As mentioned, we need a dramatic shift in our approach to conservation and that approach must center on what matters most to us – our conservation values (Leopold 1949; Winter and Koger 2004; Moore and Nelson 2010). This approach must be built upon identifying and promoting those conservation values which are broadly shared throughout our society. It must address how those values might be identified, refined, and integrated into the fabric of our communities and the nation as actions and behaviors – and subsequently norms – the ultimate reflection of an effective conservation initiative. Norms are especially important because people view a value as appropriate to the extent that others support it, especially individuals similar to themselves. And that is what a norm is – a widely-shared value.

"Setting an example is not the main means of influencing another, it is the only means."

– ALBERT EINSTEIN

This will require us all to look at conservation efforts through a different set of lenses. In place of the increasingly disconnected approach of setting conservation targets insensitive to cultural circumstances, or focused on the interests of a small but powerful sector, we need to shift our attention towards the broader conservation values of society as a whole and growing them over time to be more compatible with a living planet.

This book aims to present a comprehensive, more impactful approach. It is my hope that the book will serve as a launching pad for those with a hunger for better and more effective conservation practices and strategies.

Turning to specifics – how to achieve the effective conservation of the natural world around us – the book lays out a practical, achievable approach towards implementing conservation. This framework is built upon societal conservation values as its central element, its basic foundation. Upon this base, it then uses the prodigious advances in our understanding of human decision-making, marketing, communications, education, and other important fields – the essential building blocks that shape society – for framing societal values in a manner that communities will embrace. It then builds from there. All told, it includes nine framework elements important, if not essential, for achieving effect environmental conservation.

This approach is quite contrary to much of present conservation practice. Though periodically recognized as playing an important role, the importance of values typically receives little more than hand-waving attention other than by "human dimensions" professionals, a portion of the conservation community which remains well outside the mainstream of conservation practice (Gigliotti et al. 2009), along with a more sizable group of academicians scarcely involved in the application of conservation in any major form.

Is this proposed framework intended to replace present practice? The question should not be "either" "or." It is a matter of existing conservation initiatives re-working their programs to include these important critical elements. Obviously, some of the conservation work underway today is essential. At the same time, much of it is misdirected, achieves little, and may even be counterproductive. The approach which I propose will require a dramatic change in emphasis and the manner in which much conservation is framed and implemented. As a consequence, the training of future conservation professionals, for example, will necessitate a major overhaul. Future professionals will require a core skill-set centered on interacting with communities and working in teams rather than an emphasis on conservation biology, independent research, and thesis writing. Conservation must undertake a major shift of attention towards urban areas, landowners, and the private sector. A refocus of our perspective on what drives conservation, and who it should serve, is in order.

CHAPTER 1

TRANSFORMING OUR VIEW OF CONSERVATION

Sometimes we foresee events that transform our lives. Other times such incidents emerge out of the blue. Such was the case when Jack Turnell walked into my office a few years back.

Jack's visit was a fluke of scheduling – a time filler. Jack was in Washington, D.C. to meet with the director of the US Fish and Wildlife Service, the agency for which I worked and whose mission is to conserve wildlife and plants for the nation's people. Actually, Jack was going to do more than meet with the director. He was going to "be" the director – for a day. Jack was part of an innovative Service program entitled "Walk a Mile in My Boots." The purpose was to give some of our agency's primary constituents – among them hunters, ranchers, farmers – first-hand experience of being a government bureaucrat and, vice versa, to give Service personnel the opportunity to spend time in the shoes of a landowner who was subject to the regulatory processes our agency imposed. The intended outcome of these experiences was to create greater sensitivity to one another's perspectives and concerns.

Jack was a rancher. Apparently a big-time rancher considering he had been selected to be our acting director for a day. In any case, it was on his ranch in Wyoming that the black-footed ferret, declared extinct in 1979, had been rediscovered. For his support of various conservation efforts, including the ferret recovery program, Jack had received much deserved attention.

BLACK-FOOTED FERRET

Despite Jack's work, I saw little point to our meeting. My responsibility within the Service was international conservation, not domestic. I worked with other countries to conserve wildlife, habitats, and ecosystems around the world. These efforts ranged from saving world heritage species such as tigers, elephants, and gorillas from extinction; to enhancing the survival of migratory species, such as birds, bats, and butterflies, that fly south of the US; to the implementation of international treaties such as the Convention of Wetlands of International Importance.

Though Jack had agreed to participate in Walk a Mile in My Boots, it was clear early on in our conversation that there was no love lost between him and people of my ilk. Jack did not like bureaucrats. He did not especially like the way agencies like mine did our business – how we treated the private sector, how we addressed land management, how we went about attempting to save endangered species, and the like.

I could at least sympathize with much of what Jack had to say – many of the shortcomings conservation professionals may have here in the US are magnified in spades when practiced abroad. Indeed, some protected areas in other countries have been partially destroyed as a result of resentments generated by their being declared or managed in a careless and callous manner. All too many conservationists believe that any action performed in the name of conservation is bound to have positive consequences, with the only question being what level of success will be achieved. And all too few conservationists recognize that poorly implemented conservation initiatives, though executed with the best of intentions, have the potential not only to be total failures, but to be counterproductive.

Jack and I discussed these matters during our half hour chat, but I was left with the distinct impression he was disinterested in and impervious to my comments on how conservation might most effectively be implemented. Perhaps it was his disdain for bureaucracy. Whatever the reason, the conversation didn't seem fruitful, and I chalked it up as one of those unproductive meetings which we all experience.

The first indication that I had misinterpreted Jack's response came immediately thereafter. Following our meeting, at a larger Fish and Wildlife Service gathering, Jack made positive references to our previous conversation. I was pleasantly surprised.

Three weeks later I received a phone call.

"This is Jack. Do you remember me?"

"Oh, yes, certainly." Yes, I remembered Jack, but what could he possibly be calling about now?

"Well, I've been thinking, and I would like to do a project with you."

"A project with me?" What kind of project could he possibly have in mind doing together? I ran international conservation projects. Jack raised cattle in the American West. "What type of project do you have in mind?"

"Oh, I don't know. I was thinking you might come up with something." Huh?

"I'm sorry, but I don't quite understand...." And so the conversation went.

Though Jack had received a boatload of awards for his contributions as a private landowner to conservation, it turned out he wanted to leave a more substantive, ever-lasting legacy. He was not sure, however, what shape such a legacy should take. This legacy would be something grand – in the order of $0.5 billion to 1 billion in size – and he wanted me to draft it.

Now that's transformative!

To make a long story short, I went out to Jack's ranch. It was about 100,000 acres – and this was after it had been subdivided! Every mountain one could see from his home was part of his spread. We spent two days discussing ideas around which to build his legacy. The result was an outline for training future conservation professionals in a manner dramatically different from most such training programs today. Presently, the primary focus is on the science of wildlife and resource management. Too few recognize that resource conservation is actually about people and how people treat resources. The goal of our proposed program would be to develop talented professionals serious about conservation, but with an emphasis on the skills necessary to communicate effectively with, empathize with, and develop relationships of trust with the "Jacks" and all other citizens of our country. The importance of conservation values at various scales, particularly the community and national levels, would receive major attention. Students would learn how we develop our attitudes, values, and beliefs, how they are influenced, and how they may or may not affect our behavior and actions. Working in teams, rather than individually, would be the norm, not the exception. Such would enhance key skills of collaboration and sharing. This model program, the subject of Chapter 8, would then be replicated on a national scale to ensure its availability in every state.

What I came to recognize through my experience with Jack was how readily chasms between people can be spanned if two fundamental elements serve as the springboard for dialogue. Most important is the building of trust. The other is the seeking of commonalities – looking for areas of common interest or common values. After all, Jack was a rancher as opposite to me in nearly every way as two white, male Americans can be. I was born in Brooklyn, New York – the congested inner city. Jack was from the wide-open spaces of Wyoming. I grew up among row houses. Jack grew up on a ranch. I generally reside down one end of the political spectrum. Jack is at the other. Despite these differences, when it came to the issue of how best to achieve conservation and the skill sets conservation professionals should have, Jack and I saw eye to eye. We had similar values. When we focused on this, we made a great team.

Jack and I held contrary views regarding the reintroduction of wolves out west and their role in the environment. We had divergent opinions on the role of hunting in conservation. This did not matter. What was important was that we sought out aspects of conservation where our thinking aligned, and we made progress on those fronts. Such overlaps occur among virtually all of us. Our challenge is whether we focus on our differences, or seek areas of commonality.

Despite having already worked for decades in the conservation profession, my encounter with Jack gave me a fresh appreciation of the importance of people's values in achieving conservation goals, and I began to realize how the initiatives I managed, programs of USFWS generally, and efforts of the conservation community as a whole, ineffectively addressed this critical fact. Working with Jack prompted me to think in depth about the skills I believe conservation professionals should ideally possess. This would mean:

- Less biology and more social sciences in their training,
- Less individual research and more practical teamwork,
- Less field data analysis and more questionnaires to understand community sentiments,

- Less emphasis on individual intellectual development and more on developing trust,
- Less of a focus on understanding nature and more on empowering others to appreciate nature.

Though Jack's legacy project did not become a reality, primarily due to the economic recession of the early 2000s, the thinking that went into it, the recognition of the importance of identifying common values and of developing trust, became a foundation for this book. The more I looked, the more I came to see how essential values are in whether conservation initiatives succeed or fail. At the same time, as I explored this with my project staff, many of them with graduate degrees in animal ecology, I realized how this need was not the least bit apparent to them. It is critical that we shift our efforts as professionals, as citizens, and as a society towards values-based conservation.

Consider, for example, how much a focus on values lies at the heart of two highly successful conservation projects, the Saint Lucia Parrot and conservation of the monarch butterfly. Let's look more closely at each of these.

The Saint Lucia Parrot: How Local Pride Saved the National Bird

Saint Lucia is a small Caribbean island in the Lesser Antilles. Like its neighbors, it has a fascinating history, culture, and biogeography, but our primary interest here is its parrot known as – the Saint Lucia parrot to the parrot fancier, or the Jacquot to Saint Lucians.

Like many other Caribbean islands, Saint Lucia was colonized by Europeans approximately 500 years ago, and for most of the intervening years, it supported an agricultural society centered on the production of sugar cane. More recently, agriculture has given way to tourism as the focus of the island's economy.

Neither agriculture nor tourism is conducive to forest preservation, consequently, Saint Lucia's forests have long been decimated, leaving only small hints of their former grandeur.

By the waning decades of the 20th century, Saint Lucia's parrot, like most parrot species in the Caribbean, was in danger of extinction. There were as few as 100 birds remaining in the 1970s, and it appeared the Jacquot would soon go the way of the Carolina parakeet, North America's only representative of the parrot family, which became extinct in the 1930s. Already the Jacquot had joined a lengthy list of other parrots in steep decline the world over. In the Caribbean alone, 10 of that region's 11 surviving native parrots (not counting macaws that had died out centuries earlier) were threatened and dropping perilously in numbers. Among them also was one endemic to Puerto Rico that, by the 1970s, had become one of the rarest parrots in the world: only 13 individuals were alive in the wild.

It is not coincidental that the Saint Lucia parrot, the Puerto Rican parrot, and so many of their relatives had become endangered. Parrots have several characteristics that have been their undoing in an increasingly human-dominated world. For one, the vast majority of species nest in tree cavities. This should not be taken to mean just any tree cavity. Parrots, generally, are fairly large birds, the Jacquot being nine inches tall, and consequently require large tree cavities. As fate would have it, large trees capable of supporting adequate nest cavities also happen to be in great demand by none other than Homo sapiens. Suffice it to say that human demand for timber, combined with the clearing of forests for development, make up the single greatest threat to this family of birds.

A second characteristic of parrots, for which they suffer immeasurably, is their habit of being frugivores – they feed primarily on fruits. To make matters worse, they have the audacity to like fruits that we humans cherish – oranges, grapefruits, apples, cherries, you name it. This resulted in farmers shooting them as pests. Parrots are also pretty and easily tamed – two other characteristics that have made them vulnerable to extinction. Some can even imitate the human voice. What better creature to have as a pet and keep in a cage! Further, the very rarity of parrots has increased demand for them, as reflected in the price collectors are willing to pay for them. With characteristics such as these, it's surprising that so many parrot species continue to survive at all.

Despite these insults to Caribbean parrot populations, there have been remarkable success stories in the restoration of some of these species up and down the West Indian archipelago in relatively recent years. The Saint Lucia parrot is the most impressive case and much of the inspiration for recovery efforts on its behalf can be traced to the work of a man named Paul Butler and the staff of Saint Lucia's Forestry Department. Now, thanks primarily to the campaign they initiated in the 1970s, the Saint Lucia parrot numbers over 2000 individuals and climbing. In fact, the parrot is reverting to becoming the pest to farmers and fruit growers it likely was centuries ago – a bad thing in one way, but a very good one in another.

What happened?

What happened was that Saint Lucia developed one of the most outstanding conservation programs ever devised. What was so distinctive about the program that restored the Saint Lucia parrot was its unique focus on local values along with the thoroughness with which it was implemented.

The focus of the initiative was to place the parrot in the hearts and minds of all Saint Lucians. This is not to say that Saint Lucians did not know the parrot. Virtually all did. What they did not know was just how special their bird is.

To most young children the animals, plants, landscapes, and general circumstances under which they grow up are believed to be the same as those of all other children regardless of where they live. It is only as we grow older and learn about other places that we come to understand the similarities and differences between the place we were raised as compared to other localities. And even then, we may not recognize what's unique in our own land unless it's pointed out to us. Though all Saint Lucians knew of the Saint Lucia parrot, practically none were aware that this parrot that bears the island's name was found only on Saint Lucia and no place else in the world. It is what biologists refer to as an endemic species.

It was Paul Butler, a young Brit with a background in social marketing, who conceived of applying concepts of that field to addressing the plight of the parrot. He was visiting Saint Lucia on a university study tour and became involved with the island's Forestry Department in its parrot conservation efforts.

Paul began by making the Jacquot the "spokesperson" for a conservation campaign. Through trial and error the initiative took shape. Schoolchildren were the initial focus. The Jacquot spoke to them through a newsletter and visited them in the classroom and schoolyard. These visitations by Paul or one of the Forestry Department staff were not conducted in the traditional manner – as government officials coming to inform the kids. Innovatively, they came dressed in a make-shift parrot costume. This had a dramatic impact on young children. Teachers were provided lesson plans to help them educate their students about the parrot, its importance to Saint Lucia's cultural heritage, and its plight. This was coupled with information regarding the importance of the Jacquot's forest home for water and soil conservation on the island.

The campaign expanded to involve every school on Saint Lucia. From there it moved beyond the schools to reach out to Saint Lucian society more broadly. Every type of media on the island was utilized. Local musicians were engaged to write and sing songs about the uniqueness of the Jacquot and the importance of saving it. Sermons were written for clergy interested in addressing the issue before their congregations. Businesses were encouraged to incorporate the Jacquot on their logos.

Central to the success of the campaign was that it was not dominated by biological facts and figures about the parrot's status. Such data are of scant interest to most people other than the biological researchers who create them. Instead, the core message was that the parrot is as unique to Saint Lucia as any Saint Lucian. It had been on the island, and only that one island, for countless millennia thus making it an important, distinctive member of the Saint Lucia community. The benefit of this approach was its appeal to the deep pride local people take in their sense of place – things which make their home, or homeland, special. Pride in one's home and place of origin is a powerful value in any person or society. By portraying the Saint Lucia parrot as a unique representative of Saint Lucia, this powerful human sentiment was brought to bear in the bird's favor.

To be fair, the parrot's endemism and evolution on Saint Lucia are biological facts. The point is that heavy emphasis on biological research should not serve as the hub of most conservation initiates. It should fill a complementary and secondary role.

The campaign was transformative. Surveys conducted before and after the multi-year campaign revealed a dramatic shift in attitudes of islanders towards the parrot. Saint Lucia's Jacquot had been taken into the hearts and minds of the island's people. Islanders had learned that the parrot deserved a distinguished place in their national heritage. It was something special about which they could be proud.

This dramatic shift in attitude subsequently led to behavioral changes including a considerable increase in public support for new conservation legislation, participation in parrot counts, and visits to the forest reserve. The attitudinal changes also had an effect on Saint Lucia's political establishment, since, as is the case everywhere, politicians keep their ears close to the ground so as to detect the interests of the people they govern. Among other legislative changes, the forestry law was revised to increase penalties for violations, such as killing a parrot or intentionally poaching nests. Fines went from 48 EC$ to 5,000 EC$ – a hundred-fold increase – and one year in jail.

Increased appreciation of the parrot also led to other positive changes in what were, up until then, traditional behaviors of the populace. Keeping of parrots as house pets had been a common practice in Saint Lucia, for example. Traveling the countryside one would regularly see caged Saint Lucia parrots decorating verandas. Evidently, the procurement of pet parrots was a great threat to the species. Not only were these increasingly rare parrots being taken from the wild, but often the scarce trees with adequate nesting cavities were cut down to capture the birds. As a result of the pride campaign, Saint Lucians not only refrained from keeping parrots as pets, they even turned in dozens of pet parrots as a gesture supporting restoration of the bird. And, in one telling anecdote, a local taxi driver, upon learning the destination of his fare was the forest reserve to procure a Saint Lucia parrot for smuggling back to his home country, drove the visitor to the police station instead!

There are a number of important points to note in the exceptional success of Saint Lucia's pride campaign. Most important was the effect of the campaign's focus on people's hearts and national pride. As Saint Lucians began to recognize the uniqueness of their Jacquot, their attitudes shifted in favor of saving the parrot. This attitude shift led, in turn, to the various positive actions and behaviors described above.

A second point concerns Saint Lucia's economic status. It is often said that concern for conservation is a luxury reserved for the relatively affluent. Yet, in the 1970s, Saint Lucia's average per capita annual income was in the order of $2,000 per year as compared to $20,000 in the US at the time. Despite many economic challenges that the country faced in the provision of basic infrastructure such as housing, schools, health clinics, roads, and the like, Saint Lucians found a means to embrace their parrot and recognize the importance of conservation for their nation's future.

Third, this dramatic turn-around in attitude and behaviors was achieved with scant financial resources. The campaign to develop pride was extraordinarily cost-effective – just a few tens of thousands of dollars. It also had positive spin-offs, such as reducing habitat destruction and promoting the desire to set aside additional protected areas. Saint Lucia's prime minister actually sent a specific request for assistance in this regard to the US Fish and Wildlife Service – a rare request from such a high level of government.

SAINT LUCIA PARROT

PUERTO RICAN PARROT

Comparison of Saint Lucia and Puerto Rican Parrot Conservation Initiatives

Comparison of the campaign's achievements and cost to save the Saint Lucia parrot with that of efforts on behalf of the Puerto

Rican parrot a few islands further north is instructive. Efforts to conserve the Puerto Rican parrot may well be claimed as one of the US's wildlife success stories. At the time I began my career as a field biologist in Puerto Rico in the early 1970s, the Puerto Rican parrot population had declined to an all-time low of 13 birds in the wild. It had become one of the rarest birds in the world. Over 40 years later, prior to Hurricane Maria which devastated Puerto Rico in September 2017, the number of Puerto Rican parrots in the wild was in the order of 200 birds. Such success was the result not only of decades of extensive scientific research, management, and inter-agency collaboration, but also many tens of millions of dollars of investment. Attempts to influence human values had little if anything to do with the bird's recovery, however.

Why such different approaches between the two islands? The problems facing both the Saint Lucia and Puerto Rican parrots were quite similar: habitat destruction, poaching, nest-tree cutting, local indifference. The reason for the difference in conservation strategies was primarily that of leadership and vision. In Saint Lucia, Paul Butler developed a new vision, refined it, and demonstrated that it worked. He thought innovatively. In Puerto Rico, by contrast, a traditional approach based upon field biology was used which, despite its accomplishments, cost 10 to 100 times more than the Saint Lucia model and, in significant ways, has yet to be as successful. Most importantly, it has scarcely sought to amass the will of the Puerto Rican people behind it. Relative to Saint Lucia, pride in the island's endemic parrot is minimal. As a result, political awareness and support for the Puerto Rican parrot is far less.

One caveat. In all fairness, Puerto Rican parrot numbers in the 1970s were so low, just 13 birds, that recovery of the species built on developing national pride almost certainly would have been unsuccessful. At the same time, had parrot numbers been more substantial, similar to parrot numbers in Saint Lucia, such an approach would not have been considered as, based on personal discussions with them, it was outside the frame of reference and skill sets of Puerto Rican parrot biologists.

In a nutshell, the traditional approach applied to conservation of the Puerto Rican parrot began with research on its status: knowing its distribution, abundance, breeding ecology, and threats to its survival. Such information would be the basis upon which concrete conservation actions could then be developed. But, what was happening to the parrot population while all these data were being collected? Not surprisingly, the bird was still declining. In fact, by the time the problems facing the bird were reasonably sorted out, which arguably spanned over a decade, the Puerto Rican parrot was in such dire straits that a number of the possible actions which might have helped save it, such as trans-locating a few birds to new potential breeding grounds, were now no longer feasible. Moreover, it is believed a former poacher, hired to locate parrot nests, actually continued poaching parrots located during the study.

This is quite contrary to the experience on Saint Lucia where, as we saw with the taxi driver, poaching had become frowned upon. The most outstanding ornithologist to work on the Puerto Rican parrot, Noel Snyder, recipient of the Brewster Award from the American Ornithologists' Union for a career of outstanding research, once declared to me, "A single Paul Butler is worth a hundred of us" referring to scientific researchers. Noel was right. The renowned Jane Goodall (2015), was of equal mind, "I believe the most crucial aspect is for the local people to develop a sense of pride… and a sense of ownership."

The situation regarding the Puerto Rican parrot was not unique. In fact, its over-emphasis on research at the expense of addressing the social issues associated with conservation was, and to a large extent still is, standard procedure. A few years ago a widely respected wildlife researcher charged with conserving the endangered Hawaiian honeycreepers, a unique group of birds, bemoaned that he may well be the only researcher in the world to have documented the extinction of six species of birds. Perhaps this gentleman should have considered a different approach to doing his job.

Because the Puerto Rican parrot was already so seriously endangered when conservation actions were finally put in place, captive breeding to save the species became a necessity. Captive breeding is not cheap – it requires facilities, highly trained staff, sanitation, security, feeding, veterinary care, and the list goes on. Following decades of difficulties, the captive rearing program became successful, and Puerto Rican parrots have now been released in numbers into superior lowland habitat. The released birds are only quasi-wild, though, because they are provided food on a regular basis by aviary staff. And, because limited outreach has been done to farmers in the vicinity, it remains to be seen whether they take retribution on the parrots for the inevitable future crop damage.

Also, note the relatively minor role played by parrot research data in reversing concern about the Jacquot's decline. Reams of biological data did not change the attitudes of Saint Lucians; what worked was a few biological tidbits coupled with major appeals to their emotions and gut feelings. The biological sciences are, of course, important to conservation. They can track down the causes of decline or resurgence, they can project the likely consequences of various courses of action, and they can tell us about the sizes and trends in animal or plant populations. But when it comes to encouraging changes of heart, public support, or political action, arguments from the biological sciences alone seldom suffice.

The Saint Lucia parrot is not the prettiest bird in the world. It is not even the prettiest parrot. It does have a beautiful blue head, but as parrots go, it is hardly spectacular. Despite this, thanks to an innovative campaign, it was made into a flagship species equal in stature to the tiger in India as far as Saint Lucians were concerned. The point here is that so much of beauty is in the eye of the beholder. The power of emotion and feelings of the heart can be much more potent than that of intellect alone, or even the pocketbook.

In the mid-1990s, I had the pleasure of visiting Saint Lucia. It had been more than a decade since Paul Butler's initial pride campaign. As I checked into the hotel, I heard the characteristic squawk of a parrot and raised my head. Noting my curiosity, the young woman registering me smiled broadly and said, "That's the Jacquot, Amazona versicolor. It's our national bird." Now *that* is getting conservation into people's hearts where it belongs!

The transformation in parrot conservation was so successful in Saint Lucia that the conservation organization Rare adopted the Pride Campaign as its signature focus and, with Paul Butler continuing to refine its delivery, initiated similar campaigns on other islands of the Caribbean where they also proved highly successful. Beginning with the Saint Vincent and imperial parrots, endemic to Saint Vincent and Dominica respectively, campaigns expanded to other islands using other species as flagships, such as the Jamaican giant swallowtail, the largest butterfly in the Western Hemisphere. To date, many hundreds of pride campaigns have been initiated by Rare around the world, their success continuing to be exemplary.

Pride campaigns are one powerful tool for the conservation toolbox. They are a good example of a **Cornerstone Initiative,** one of our framework elements discussed in Chapter 8. There are others. Our second values-related example involves a totally different approach. It focuses on addressing a community's basic needs while gradually, over time, connecting those needs to a concern for conservation.

MONARCH BUTTERFLY

The Monarch of Migration: Conserving the Monarch Butterfly

High on the mountain massifs of central Mexico, amidst the swirling clouds and majestic oyamel fir forests, resides, during the full extent of northern winters, an extraordinary insect whose presence, during summer, most in the United States take for granted in our own backyards. This is the monarch butterfly, a species well known, but whose phenomenal life history draws much less attention. This beautiful orange and black butterfly undertakes an annual migration from eastern North America down to central Mexico where it congregates in huge colonies often numbering 50 million or more individuals. Here, dangling from the fir trees in spectacular densities, the butterflies hang in torpor until spring arrives and their northward migration begins. To make this migration even more remarkable, the butterflies actually take several generations to return to their northern climes, as far north as Canada, from which their ancestors originally departed.

Tragically, the oyamel fir forests of the monarchs happen to belong to the most endangered forest type in Mexico, and the only forest type the monarchs find suitable. Only two percent of these native forests survive, the remainder having been cleared for various purposes, particularly to accommodate subsistence agriculture.

In 1986, with pressure from US-based conservation organizations, the president of Mexico decreed 60 square miles of oyamel forest as a Monarch Butterfly Special Biosphere Reserve. Though a victory for conservationists, this was hardly a victory for the butterfly. In Mexico, unlike the United States, national parks and protected areas are declared without shifting ownership of the land from private hands to those of the government. The result of the presidential edict was, thus, to impose additional constraints on land use by already impoverished peasants living within the newly designated reserve without consultation with them on the matter and without compensation. This, unquestionably, was not well received and soon led to such resentment among local communities in the reserve that some individuals began to cut and burn the forest in protest.

That was one approach to conserving the monarch in Mexico – US-born, culturally insensitive, politically naive. There have been others. Some of these too have failed, though not as disastrously as the 1986 decree. There has also been a masterful success story and it bears a closer look.

Because of the failures of previous efforts, communities in and around the monarch reserve had become highly suspicious of conservationists. Why shouldn't they be? Recognizing this, and that monarch butterfly conservation had to be built around the wants and needs of the people of the area, two young women, Guadalupe del Rio and Ana Maria Muñiz, subsequent founders of the group Alternare ("Alternatives" in English), in the mid-1990s, sought a way to engage these now suspicious communities. Being from Mexico City, and scarcely expert in rural agriculture, Lupita and Anita, as known to their friends, realized a unique mechanism was needed to reach out to the local campesinos. The solution: the women engaged Gabriel, a peasant farmer from an adjacent region of Mexico, and an expert in rural agriculture who was willing to transplant his family to the vicinity of the monarch reserve and bring new skills to the local people. Such skills were needed because many of the communities surrounding the reserve were composed of families originally from elsewhere in Mexico. These individuals had been provided land up in the mountains following the Mexican Revolution (1910-1920) and, being displaced to a totally new environment, had no idea how to manage the land productively.

Gabriel began by teaching the benefits of proper composting. He then expanded his instruction to address soil erosion and promote better crop rotations. Building forest-friendly houses out of adobe rather than wood was among the skills Gabriel transferred to the local communities. The use of more efficient wood-burning stoves was another. Though it took a while, years in fact, Alternare gradually gained ever-increasing levels of local trust. They achieved this by focusing on their core values of respect, responsibility, honesty, and commitment. And they listened. They listened to the communities and responded to their needs. Extremely important was their early success in enabling the communities to live more prosperously off the land. This helped open the eyes of many skeptics. Over time, with a scant budget of $100,000 per year, Alternare engaged the most successful local campesinos as "*promotores*" – informal extension agents – who were better able to develop a rapport with neighbors and adjacent communities. As of 2016 there are over 51 such *promotores*. Many of the most avid and effective *promotores* were women, an important empowering element of this initiative.

How did these efforts contribute to conserving the habitat of monarch butterflies? Crops need water. The local people learned how intact forests help capture water which can later be used for crops. Cooking requires wood thus, simple, efficient stoves helped save on the need for timber cutting. Typical houses required timber, but converting to adobe, made from clay, dramatically reduced this need, plus made for a much more long-lived home. Learning about fire control and prevention reduced forest loss, as did the creation of tree plantations. As a consequence of these new insights and livelihood changes, the demand for timber from the forest on which the butterfly depended was decidedly reduced.

Alternare deliberately refrained from focusing on butterflies in the course of this work. Though butterflies were their long-term interest, Alternare recognized that building trust by focusing on the community's greatest needs was a necessary first-step. Without mutual trust, community engagement in conservation of the butterfly would be impossible. To address this, emphasis was placed entirely on improving livelihoods which, in turn, resulted in both the wiser use of resources and a greater appreciation of the forest and the benefits it provides. Over time, through expansion of more effective living and farming strategies, the image of the forest, the reserve, and the butterfly shifted from being a constraint on the livelihoods of the people to something that was now understood and cherished. The power of this shift in community attitudes and perceptions was dramatically demonstrated when, as a result of increased illegal logging by outside entities wishing to make a profit off the wood, communities with which Alternare collaborated actually sent contingents of armed guards into the forest to protect the trees from vandalism.

Alternare's approach was slow. It was tedious. It did not focus on the butterflies directly, but rather on their habitat, which became as important to the communities surrounding the forest as it was to the butterflies themselves. Through over 500 workshops and tireless engagement with families and communities for over two decades, the value of the forest, the importance of reserve status in protecting the forest, and the importance of the butterfly in a decision to create the reserve, has been reversed in the hearts of local people from an attitude of disrespect for the butterfly to one of appreciation. The result: a level of effective conservation unachievable by more conventionally accepted approaches such as (1) the failed 1986 decree which simply set aside the monarch reserve, but led to degradation of the butterfly's habitat, not its preservation; or (2) by foreign scientists who monitor butterfly numbers, sound the alarm when numbers are down or the forest is cut, but are short on practical solutions when it comes to how best to address the situation. To the contrary, Alternare found a long-term solution that has worked.

Why so little progress?

Why then, after so many decades, and thousands of conservation projects implemented at home and abroad, have not these approaches of reaching out to the hearts of people as was achieved in the examples from Mexico and Saint Lucia, and of seeking out common ground – common values – and of developing trust as I experienced with Jack Turnell – why have such approaches not taken hold? There are seven key elements in answer to this important question because they are so central to the problem before us.

Let's look at each of these important obstacles to a values-based approach to conservation:

Obstacles to values-based conservation:

1. Changing Context
2. Change Is Hard
3. Belief Money Solves All Problems
4. Simple Solutions Rarely Work
5. Society Is Outcome Oriented
6. Changing Values Is Too Slow
7. Values Don't Always Match Behaviors

1: Changing Context

One major cause of this disconnect – a focus on animals and plants outside of the context of the human environment around them – probably derives from the early days of wildlife conservation, the late 19th and early 20th centuries, when the study of animals as compared to the practice of conserving them, were not very far apart as disciplines. After all, during that period you could conserve wildlife in areas where scanty human populations were a non-factor in this process. Over time, however, our burgeoning human population has increasingly impinged on wildlife causing increased conflict with it. Fundamentally, the vast expansion of people into habitats that previously were nearly uninhabitable, deserts, and the arctic for example, has made conservation almost entirely a "people problem" not a biological one. Over time, therefore, the study of wildlife versus the conservation of it, have diverged – substantially. Because this shift has been a slow, inconspicuous occurrence, it has not been well recognized nor have the necessary adjustments been made. Successful strategies of years ago have not adapted to changed times. Conservation of our nation's first national park – Yellowstone – is an excellent example of this. When Yellowstone was designated a national park back in 1872, the American West was still decidedly wild. Five years after the park's designation the Nez Perce Indian tribe, in flight from the US cavalry, fled through the center of Yellowstone and had several hostile encounters with park tourists. What better example of how lightly inhabited and remote parks were. But look at Yellowstone today! It is surrounded by towns, ranches, and farms. Approximately three and a half million people visit the park annually. Does it not now make sense that conservation practice should have undertaken a dramatic shift towards working with people?

2: Change Is Hard

A second key element, as in all human endeavors, is that change is always difficult. The status quo may be the strongest force influencing everything we do. It is an anchor against change. When reaching a set goal requires a new approach, addressing the status quo will always be a major challenge.

3: Belief That Money Solves Problems

A third element is that many people, professional or not, would argue the problem of achieving effective conservation is not about the approach – a focus on species rather than values – but is a matter of money. They believe if funding could be considerably ramped up, everything would be fine. I doubt that. I believe a fundamental reason there is not more funding is because the present approaches for delivering conservation have limited capacity to generate the social and political support they need. Scant funds are a consequence of this – not a good sign. We shall see in the following chapters that the availability of funds apparently is of minor consequence when compared to the importance of a society's values system. We have already seen, in the cases of the Saint Lucia parrot and monarch butterfly, the notable success of conservation projects which focused on changing attitudes and values that achieved success with modest amounts of funding. Success hinges more often upon the approach used, not the money spent.

4: Simple Solutions Rarely Work

Another reason a more values-oriented approach hasn't taken hold is our desire for simple solutions. As a consequence, we frequently fail to recognize that what appears the most direct route to a solution isn't always the most effective in the long run. If we wish to conserve wildlife, the thinking goes, then let's focus directly on wildlife. Likewise, if we want to conserve habitat, then let's find the easiest way to set it aside. In that context, while a number of conservationists make the case that developing a sense of shared values among our populace may be a worthy aim, it should take a back seat to addressing, through land purchase and other actions, the urgent problems we face of conserving species, saving habitats, reversing climate change, and the like. Though, these goals, as lofty and urgent as they may be, cannot be effectively achieved without our populace first having a set of widely shared conservation values in place. Focusing on people's positive conservation-oriented values and making their identification and adoption a major priority is essential if conservation is to advance, as it desperately must.

Aldo Leopold (1949), the renowned American conservationist of the mid-20th century, recognized this centrality of values. "I think we have here the root of the problem," he wrote. "What conservation education must build is an ethical underpinning for land economics and a universal curiosity to understand the land mechanism. Conservation may then follow."

Many in the professional conservation community have never taken to heart Leopold's insight. They have chosen, instead, to do what is most comfortable – to do what they are trained to do – to study and manage wildlife and their habitats – or whatever other resources they are charged with managing. They remain driven by the idea that their core leadership role revolves simply around the science of studying wildlife, not of using the social sciences to work with people and address their basic conservation values.

Case in point: In 2020 I attended a "listening session" held by the Massachusetts Division of Fisheries and Wildlife devoted to hearing what the people of Cape Cod thought about a coyote killing contest sponsored by a local gun and outfitting shop. The contest offered prizes for the largest coyote killed, the most coyotes killed, and the like. One local outdoor writer equated such a contest to any other – a beauty contest for example.

The vast majority of the nearly 200 meeting attendees thought otherwise. Most considered it barbarous, inhumane, and cruel killing for fun. And what did the Massachusetts Division of Fish and Game think? Well, for one thing, the Division never would have even held the meeting had it not been pressed to by local politicians. Such contests were perfectly legal in the state because coyotes are not considered game animals and so are subject to extremely loose hunting restrictions. The Division chose to usurp half of the meeting with nearly an hour-long presentation on coyote biology because that is all it was concerned about – that the contest was not threatening coyote numbers in the state. This was its sole yardstick.

The contrast in values was stark. The local populace overwhelmingly wanted the unnecessary killing stopped. The Division of Fisheries and Wildlife cared solely about applying biological data to the law. And oh – just what law is this? A law so obsolete it still contains terms such as "varmint" and "vermin" and treats "predators" as pests instead of important natural features of the environment. Such derogatory terms have a strong influence on how we view these animals and the actions we take to protect, or destroy them (Bekoff 2014). It was quite an experience listening to a boat load of up-to-date scientific data being applied to justify the continuity of long obsolete and inhumane practices under an arcane law that ignores the values of the majority of the Cape's, and the state's, citizens.

The premise of the overwhelming majority of resource managers with whom I have interacted over decades, particularly in the US and other nations of the developed world, is that no meaningful conservation can be achieved until we have a reasonable idea of the status and distribution of the species and habitats we wish to conserve. From there, they anticipate some top-down intervention that will implement their professional recommendations. This fundamental misconception is the underlying flaw of much conservation as practiced today and, as well, for many decades in the past. It is simply not widely acknowledged that the science associated with the study of wildlife to achieve conservation is a tool, one of many. It is simply not believed that values are the basic foundation of conservation without which no effective conservation edifice can be constructed. Commented on by Dwight Barry and Max Oelschlaeger (1996), "To pretend that the acquisition of "positive knowledge" alone will avert mass extinctions is misguided." The facts that people use to make decisions are in many instances of minimal influence in their decision-making process (Zaltman 2003; Akerlof and Kennedy 2013).

The most stark example of this ill-focused approach is well illustrated by a meeting I attended between the director of the US Fish and Wildlife Service and his Mexican counterpart held in the early 2000s. Mexico's director had explained how for years his country had applied the US approach to enforcing wildlife law and had failed. They found that a heavy focus on going around and arresting violators was counterproductive. Only when Mexico shifted to having its law enforcement personnel engage local people and focus on informing them about the relevance of wildlife law and how it fit into each community's values, did they have any success. This had resulted in Mexico integrating cultural values as an important component of all the country's conservation efforts. I found the presentation inspiring. Mexico had figured it out. The US director's response? "That's regrettable. In the US we make our decisions based solely on biology. Period!"

Is this position so unusual for the US? Not really, as environmental philosopher Bryan Norton (2005) points out, a major criticism of the Environmental Protection Agency was that scientists there "consider discussions of values to be specifically forbidden topics of conversation. Norton goes further to insist that "attempts to separate factual information from value judgements have been the root of much miscommunication and dysfunctionality in environmental discourse."

5: Society Is Outcome Oriented

A fifth point is that society is outcome oriented. It wants to know the bottom line. As a result, the conservation community, too, is increasingly focused on a measurable final product. And, as far as wildlife conservation is concerned, the values people possess are not a final product but perhaps a tool to get to that goal – the goal being the number of breeding ducks we can produce or acreage of deer habitat we can set aside. I would not be the least bit surprised if the number of wetlands acres reputed to be conserved through conservation grants far exceeds the total acreage of wetlands actually existing in the US. Why? When simplistic goals are set, the entities responsible for achieving those goals interpret their efforts in the light most favorable to themselves.

As an example, I recall reviewing a report claiming that 20,000 acres of habitat had been conserved but, when I looked deeper, the claim was based upon the fact that a brochure costing $5 thousand highlighting the values of this acreage had been funded by the reporting entity and so they chose to claim the land as "conserved." Though "goal-setters" may be satisfied, close scrutiny will typically demonstrate the goals have not been reached. Such is likely the case in most fields of endeavor where simple measurables are applied to complex problems.

6: Changing Values Is Too Slow

A sixth and frequently voiced objection to a focus on values is that the process is much too slow. The species we desire to save will be long gone before people's attitudes are changed, they argue. Let's consider three responses to this objection. First, had the conservation community begun addressing societal conservation values when the problem was first recognized, we might now be in the second or third generation of broad conservation thinking and many of our present dilemmas might not exist. Second, it is now well documented – views regarding cigarette smoking being an excellent example – that if children's attitudes are changed, their parents' attitudes may shift as well. Parents are affected more by their children than probably any other factor. Helping children develop a consciousness of conservation and its importance serves to influence attitudes and behaviors not only of their generation, but also that of their parents, leading in some cases to relatively rapid changes in, for example, recycling practices. Third, choosing an alternative approach just because it is easier or quicker, if it does not get us to the intended goal, is no alternative at all.

7: Values Don't Always Match Behaviors

A final objection, recent in nature and voiced by some

environmental sociologists, is that values and attitudes are not necessarily reflected in an individual's behaviors and actions (see Heberlein for a comprehensive discussion of this concern). Though this may be the case, it is likewise true that values and attitudes are the necessary underpinnings for positive conservation behaviors and societal norms. As a result, our ultimate challenge is to build norms around our society's conservation values. As Herberlein (2012) points out, "They are the key to changing environmental behaviors."

Seeking a New Path Forward

We are fortunate. There are solutions to every community's, nation's, and the globe's, conservation woes. We can reverse the declines of many species. We can dramatically reduce the threat of climate change. We can address the problems of desertification, water shortage, acid rain, ozone depletion, and deforestation. All the negative trends we see today can be reversed. But this can happen only if we are smart regarding how we go about the process – only if we identify the best tools available to us and use them wisely and effectively. My remarkable experience with Jack Turnell, the saving of the Saint Lucia parrot, and the conservation of the forest home of the monarch butterfly are indicative of the critical importance of values, trust, and people's hearts in achieving conservation. In the next chapter we look at the importance of values from a quite different perspective. We compare conservation achievements and practices in two dramatically different countries and see what we can learn from the results.

CHAPTER 2

COMPARISON: WILDLIFE IN THE U.S. AND INDIA – STATUS OF FLAGSHIP SPECIES

The American bison has long been a symbol of our country's heartland and, arguably, the nation itself. The bison's fate over recent centuries is symbolic as well – of both the good and the bad in the legacy of American wildlife conservation. Once present in the tens of millions and spanning nearly every site outside of New England, westward expansion and slaughter on a vast scale diminished bison herds to scarcely a few hundred individuals by the late 19th century. More recently, concerted efforts have made something of a comeback possible: today, in the order of 30,000 bison roam unfenced in the wild (Smithsonian Institution – National Zoo website).

AMERICAN BISON

Such is the story of much of America's wildlife. Abundant during pre-colonial and colonial times, devastated during the raw, rapid development of the East and "taming" of the West, and subsequently "restored" to varying degrees in the 20th and early 21st centuries. I put "restored" in quotation marks because for many species, like the bison, the numbers today, while typically higher than a century or so ago, are paltry compared to the pre-colonial past. To boot, upon experiencing a modicum of recovery so that they are no longer endangered, this species, among others, is then hunted once more for any of a number of pernicious reasons.

The conservation history of the whooping crane, North America's tallest bird and symbolic of avian wildness, has also proven quite dramatic. Formerly widespread in eastern mid-continental prairies from Alberta, Canada to the southeastern United States and northern Mexico, and believed to number between 10,000 to 20,000 birds, the impact of over-hunting, collecting, and particularly destruction of its prairie wetland habitat for conversion to agriculture, resulted in dramatic declines through the 1800s and early 1900s. By 1941 the entire population of this noble species numbered a mere 15 birds. Today that core flock has increased to over 500 individuals with incipient populations in various stages of being artificially established at localities in which the bugle of the whooper has been absent for more than a century (National Wildlife Federation website; International Crane Foundation website).

The list goes on. Raptors such as the bald eagle, osprey, and peregrine falcon, alarmingly reduced in numbers by the effects of widespread application of DDT in the late 1940s and the 1950s, have, with the help of extensive conservation efforts, recovered dramatically. In 2011, with nearly 10,000 pairs breeding in the lower 48 states, the bald eagle was removed from the federal endangered species list. This increase was from a low of 417 pairs in 1963 (US Fish and Wildlife Service website). The wild turkey, discussed earlier, abundant during pre-colonial times, was decimated following European settlement but has recovered spectacularly.

Many ducks and geese have recovered from over-hunting as well.

Conservation of wildlife in the United States clearly has had some impressive successes over the past century or so. What accounts for this reversal of fortune for some of the country's wildlife? In 2001 Valerius Geist and others postulated seven fundamental tenets of conservation practice that they believed were behind this shift. Collectively, these tenets became widely accepted as the North American Model of Wildlife Conservation. As summarized by Michael P. Nelson and his coauthors, the "literature about it has grown, professional organizations have endorsed it, institutions have developed curricula to teach it, state agencies have built it into their strategic plans, sessions at professional meetings have focused on explaining it, and an entire issue of The Wildlife Professional was devoted to it" (Nelson et al. 2011). Geist (2006), among others, has even claimed that development of the Model represents "probably the greatest environmental achievement of the 20th century." As such, the Model has been widely proposed for exportation and replication around the globe.

The seven tenets of the North American Model of Wildlife Conservation

1. Wildlife is a public trust resource: All animals in the wild are by law owned jointly by the people, and the government serves as caretaker.

2. Markets for the commercial sale of wildlife are eliminated: Specifically, wild meat is not to be sold, though sale of parts such as furs and skins is acceptable.

3. The uses of wildlife are established by law: Laws and regulations pertaining to wildlife are made through the democratic process.

4. Wildlife can only be killed for a legitimate purpose: wildlife is not to be killed strictly for fun or to be used for financial profit. Game killed should not be wasted.

5. Wildlife is considered an international resource: The federal government coordinates with other countries to manage shared species.

6. Science is the proper tool to inform wildlife policy: Management decisions should be based upon the best available science.

7. The opportunity to hunt is available to all citizens, not just an elite few.

Does the North American Model of Wildlife Conservation truly explain the turn for the better of wildlife conservation in the United States? Is it accurate? Answering these questions in depth here would divert our discussion, so I offer only a few cursory remarks. Frankly, the substance of the Model appears as flawed as its title. It so happens that Mexico is, by all accounts, part of North America, yet this Model in no way reflects that nation's unique and decidedly different conservation history. So is Canada, though in conservation practice it closely mirrors the United States. Failing to honor these geographic and cultural facts reflects a very regrettable and long-standing bias in American conservation – the presumption that the US is representative of our continent and that the two other nations which cohabit it are somehow less relevant. That conceit has resulted in all sorts of failed cross-border relations, conservation and otherwise, but these, as fascinating as they are, fall beyond the scope of our discussion. Suffice it to say that the North American Model is purely a US construct.

Moreover, the Model captures only one strand, albeit a dominant one, of the conservation movement, that driven by hunters and their allies. There are other strands. Among these are the awe of nature inspired by John Muir, the land ethic espoused by Aldo Leopold, and the vision of a healthy environment advocated by Rachel Carlson, that also have contributed considerably to the fabric of US conservation – though it would seem all have been in decline in recent years. These additional threads wove important new values into the fabric of conservation. Understanding US conservation history without them is near impossible.

One of its seven claims is that the stoppage of market hunting was an important element in stopping wildlife declines in the United States. Well, based upon the above discussion, that is certainly true. At the same time, is it not equally obvious that had the first European immigrants had a more earth-conscious mindset, the calamity of over-hunting would never have occurred in the first place?

Another of the Seven Sisters – the endearing moniker given to the seven tenets of the North American Model – pertains to the importance of passing key legislation to protect species, legislation such as the Lacey Act of 1900, which made market-hunting illegal nationwide, and Migratory Bird Conservation Act of 1918. Under the circumstance, no one would argue that such legislation was not essential. But we should reflect and ask: Why was it necessary? It was necessary to redress cultural values that were anathema to a sustainable ecosystem. Is it not preferable that such laws not prove necessary to begin with because a culture lives in harmony with the environment?

Michael Nelson and others (2011) in *"An Inadequate Construct: North American Model: What's Flawed, What's Missing, What's Needed"* offer additional commentary on the model's failings.

Despite the litany of shortcomings, the North American Model remains a holy doctrine in many states and hunter-oriented conservation groups.

Other important questions derive from our discussion of the North American Model. Should the Model be exported abroad? Given the inadequacies identified, I would suggest not. Has conservation success in the US been greater than in any other country? Now that is an intriguing proposition worthy of deeper examination.

In 2010 I attended a conference of conservation professionals focused on the North American Model in which the majority of participants voiced sentiments highly supportive of it. Intrigued, I inquired whether the North American Model had ever been compared to that of any other country. Apparently not: the presenters, renowned experts on the subject, expressed a desire to make such comparisons but, in their words, "had not gotten around to that yet." This response was not the least bit surprising since it was evident by the extent to which the North American Model was held in such high repute, the experts felt there was little value in considering other models that could not possibly stack up to the North American version. Despite this sentiment, making such cross-cultural comparisons, in my view, could reveal a great deal about the foundations of wildlife conservation and possible limits of the North American Model as a template for explaining the secrets of conservation success in the United States and elsewhere.

Particularly intriguing, as regards such a comparison, is the case of India, which we shall soon connect to the revoyage of the *Mayflower.* India of course differs from the United States in many ways – in history, culture, religion, economic development, and location among others. But at the same time, there are some important similarities, particularly in the general nature of India's wildlife and that nation's rapidly expanding development. This combination of differences and similarities make India an excellent country for comparison from which, it turns out, we can learn a vast amount. So, let us do just that.

Comparing the US and India

The contiguous US is about two and one-half times the size of India – 3.1 million square miles as compared to 1.3 million square miles for India. Its human population is 332 million while that of India is 1.4 billion, over four times that of the United States. These data alone reveal an important demographic – that on average every square mile of land in India is approximately 10 times more heavily populated by humans than an equivalent amount of land in the United States – in the order of 1,000 people on a square mile of land as compared to nearly 100. The consequences of this are many – all clearly negative for India. Most obvious and important is that India's people historically have lived in much closer association with their land's wildlife. The potential for human-wildlife interaction, incompatibility, and conflict has always been, therefore, much greater.

India has some spectacular wildlife species, among them the tiger, lion, leopard, cheetah, Asian elephant, Indian one-horned rhinoceros, gray wolf, king cobra, and various crocodiles. How have they fared – particularly the major mammalian predators among them – compared to their counterparts in the contiguous 48 US states?

Wildlife Conservation in the US

In the US the major mammalian predators in recent centuries have been the gray and red wolves, the grizzly bear, and the mountain lion. The overarching story here until well into the 20th century was one of relentless slaughter. This was followed, in the most severe cases, by limited attempts at restoration. Here's a look species by species.

Gray Wolf

The first of these, the gray wolf, whose range in pre-colonial times spanned nearly the entire country except the southeast, is widely agreed to have been virtually exterminated in the lower 48 states by the mid-1930s. Its extirpation was driven by various factors including fear of attack on humans, concern for livestock – particularly cattle, reputed competition with sportsmen for elk and deer, and for their pelts. Persecution began in earnest in the US as early as 1630 when the Massachusetts Bay Colony offered a wolf bounty (International Wolf Center website). Two-hundred and fifty years later, during a mere eight years from 1880 to 1887, US government bounty hunters alone are reported to have killed 385,000 wolves. This is apart from the massive extirpation efforts sponsored by the livestock industry and state agencies. All totaled, in the order of 800,000 wolves were killed during this short period (ibid.). In 1931, Congress passed the Animal Damage Control Act, giving the Secretary of Agriculture broad authority to expand "the destruction of mountain lions, wolves, coyotes, bobcats, prairie dogs, gophers, ground squirrels, jackrabbits, and other animals injurious to agriculture, horticulture, forestry, husbandry, game, or domestic animals, or that carried disease."

Wolves were poisoned, shot, trapped, snared, and tortured by being burned to death, dismembered by dogs, having their jaws wired shut, and being inoculated with mange. Over half a century after its disappearance, the federal government, in 1995, began to reverse the situation by reintroducing the gray wolf into Yellowstone National Park. The program proved so successful, from a biological perspective, that the gray wolf population in that region now numbers nearly 2,000 individuals (US Fish and Wildlife Service website). At the same time, a separate population in the western Great Lakes states expanded to 3,700 individuals (ibid.) As a consequence, in 2012 the federal government delisted these two populations of the gray wolf from the Endangered Species Act thus returning its management to the states. Several western and midwestern states have reinstituted wolf hunting. In August, 2021, Montana expanded wolf hunting to include the use of baits, traps, and snares for strangling the animals.

Red Wolf

The red wolf, far less familiar to the public, resided primarily in the southeastern and mid-Atlantic states. It too succumbed to human persecution in the same manner and for the same reasons as did its larger gray cousin. By 1970 fewer than 100 individuals survived and its range was confined to coastal Louisiana and Texas (Red Wolf Coalition website). When these animals were subsequently captured, it was discovered that only 14 were pure-blooded red wolves; the rest had hybridized either with dogs, coyotes, or gray wolves. Gone from the wild, the pure-blooded captive animals were bred in zoos to build up the species' numbers. Then, as in the case of the gray wolf, the federal government began a reintroduction program. The reintroduction, begun in North Carolina in 1987, was expanded to the Great Smoky Mountains and islands off the coasts of several southeastern states. By 2016, red wolves in North Carolina had declined to 45-60 individuals, the decline due primarily to animals being shot (Hinton et al. 2017). In 2018, the US Fish and Wildlife Service proposed hunting of these remaining wolves be allowed on private lands. Taken to court, it was found that the USFWS had violated its congressional mandate to protect the red wolf and that it had no authority to allow hunting of the remaining animals (Fears 2018). To date, there has been no successful reintroduction of the red wolf and there remain only 20 or so wild individuals in coastal North Carolina (US Fish and Wildlife Service website).

Grizzly Bear

The grizzly bear has fared only slightly better than the wolves. Between 1850 and 1920, this imposing predator had been eliminated from 95 percent of its range, leaving a scant number of individuals scattered in the remote mountains of Montana, Idaho, and Washington, down from an estimated population of 50,000 (https://defenders.org). It declined even further due to unregulated hunting up until 1975 when it was listed as threatened in the lower 48 states under the Endangered Species Act. Considering that this bear was formerly abundant throughout much of the western plains, Rocky Mountains, and Pacific coast, the range reduction was estimated at 98 percent. In 2017, the US Fish and Wildlife

Service, arguing that the bear had recovered, removed it from the federal endangered species list. That decision was met with several lawsuits, which resulted in a court order in 2018 requiring USFWS to relist the bear. Today, between 1,200 to 1,400 grizzly bears roam free in the lower 48 states (Interagency Grizzly Bear Committee website).

Mountain Lion

In the case of the mountain lion, which formerly ranged through all 48 contiguous states, extirpation occurred virtually throughout the eastern and central portion of the country, due primarily to it being considered a threat to livestock and a competitor with sportsmen for deer. As with the other major predators, bounties were placed on its head, the first being in California as early as the 1500s, with others extending to as recently as the 1960s. Despite being eliminated from all but 12 states, and having its numbers dramatically reduced where it survived, this large cat maintained a foothold in the Rocky Mountains and along portions of the Pacific coast. A relic population of approximately 20 animals survived in Florida. Presently, the mountain lion's status in the wild has improved with numbers estimated in the order of 30,000 animals including approximately 125 in Florida (Mountain Lion Foundation website; Defenders of Wildlife website).

These various predators owe their near demise in the United States mainly to their being just that – predators. In many areas, particularly the plains, their habitat was dramatically altered, but overall there is wide agreement that the conflict between these species and humans was of primary significance. These animals were generally not a direct threat to humans themselves. The wolves and mountain lions in particular were perceived as a threat to livestock, primarily cattle and sheep, and few attempts were made to learn to live with such threats. It also did not help matters that one of the major prey items of the gray wolf is the elk, an animal highly sought after by hunters. Likewise, another favorite of hunters, deer, fall prey to all four of these predators. Between ranchers and hunters, North America's major predators had formidable detractors and paid a high price for that wrath. Through intensive conservation efforts, all but the red wolf have recovered to a small extent.

The United States' Largest Herbivore

We've already seen, too, that campaigns that decimated some species weren't always confined to predators. America's largest herbivore, the bison, could hardly have come any closer to extinction than it did, as discussed at the beginning of this chapter. And the reasons for its demise scarcely involve the animal being a threat to humans. Instead, the decline resulted primarily from commercial and recreational hunting. Bison were even slaughtered to deprive indigenous peoples of their traditional food source so as to drive them to capitulate. This was a practice supported by the US military, coarsely stated by Colonel Dodge in 1867, "Every buffalo dead is an Indian gone" (Indian Country Today website).

Wildlife Conservation in India

Let's shift our attention to the fate of mammalian predators in India and to that country's largest herbivore. Quite a few predators to consider here – tiger, lion, leopard, cheetah, and, interestingly enough, the gray wolf once again. (I am omitting discussion of the snow leopard and clouded leopard due to their historically restricted natural ranges in India.)

Tiger

Tigers formerly ranged throughout Asia up into much of eastern Russia, south to the islands of Indonesia, and west to Turkey. Numbers were in the hundreds of thousands (Wildlife Conservation Society website). Present estimates place the world tiger population at close to 3,900 animals in the wild (World Wildlife Fund website), probably less than one percent of the tiger's former numbers. Of the world's remaining tigers, three-quarters – close to 3,000 individuals – survive in India (Wildlife Institute of India website).

India's tigers are scattered around the country, most in nature reserves, none bounded by fences. Remarkably, it is estimated that over four million tribespeople live within the borders of India's protected forests and reserves and many millions more live in their vicinity (BBC News website). Additionally, between 30 and 40 percent of India's tigers live outside of tiger reserves or national parks, bringing tigers in close proximity to millions more humans (Jairam Ramesh, former Indian Minister of the Environment, personal comment).

The threat tigers pose both to humans and their livestock is orders of magnitude greater than all four of the top predator species in the US combined. Between 1876 and 1912, at the same time the US was systematically eliminating its top predators, tigers are reported to have killed over 30,000 people in India (Statistical Abstracts). In 1911, the Champawat tigress alone reportedly killed 436 people in the northern state of Uttarakhand (Facts and Details website). In 2019, a minimum of 95 persons were killed by tigers in India (The Hindu website).

Just consider the dislike America's ranchers have for the wolf, an animal that despite its periodic predation on livestock is scarcely a threat to human lives. Had an animal as dangerous to humans as the tiger ranged the American West at the time of our nation manifesting its destiny in the 19th century, we can only imagine the slaughter that would have transpired. Also worth noting is that as recently as 1920, when India was already more than 5 times more densely populated than the US is today, the tiger population was believed to number in the order of 100,000 animals – a very healthy population indeed (Wildlife Conservation Society website).

Despite the continuing occasional loss of human life to tigers in India, resentment of the tiger's attacks on humans and their domesticated animals is far from being the major threat to Indian tiger survival. As a case in point, a local woman was killed in a tiger reserve I was visiting in 2023. She and two friends had entered the reserve to collect Tendu leaves which are profitably sold to wrap tobacco into beedi – thin, local cigarettes. Was there an outcry? No. Local communities well understood the risk taken when entering a tiger's domain and that this woman had entered by her own choice. The tiger had no fault and was left alone.

Rather, the greatest threats to tigers in India today are destruction and disturbance of their habitat, as well as the illegal killing of tigers for international trade, tiger parts being in enormous demand reputedly for Chinese traditional medicine. If it were not for the poaching for foreign markets, tiger numbers in India likely would be dramatically higher than at present.

Leopard

The leopard is no laggard as a pest to humans. It is notorious for taking livestock, particularly goats, which are prized domestic animals in India. Notedly, individual leopards, from time to time, have become serious man-eaters. In the early 20th century, from 1918 to 1926, what was known as the man-eating leopard of Rudraprayag was documented to have killed no fewer than 125 humans before being shot by the famous hunter Jim Corbett (Corbett 1947).

Despite these threats, reduced today but scarcely eliminated, the leopard remains well established in India, its population standing in the order of 13,000 animals (International Union for Conservation of Nature website). Its numbers, too, would likely be substantially higher were it not for the impact of poaching for the international trade market primarily as a substitute for the tiger in Chinese traditional medicine, but also for its decorative pelt.

Lion

India sustains the only wild population of lions outside of Africa. This population, numbering approximately 400 animals, survives in the Gir National Park and Lion Sanctuary and has been very low for years. In recent years, it has doubled from its all-time low of 180 individuals (Singh et al. 2011). Formerly ranging from the Middle East all the way to eastern

India, the lion was extirpated country by country, to the point that it only survives in India. The decline of the lion in India was driven by hunting as much or more by the colonial British who considered it the most aristocratic of sports than by Indian royalty. Shooting guns, at animals or otherwise, was not an Indian pastime. To illustrate: During the Indian Rebellion of 1857, a single British officer shot 300 lions (ibid.). At that same time, and in stark contrast to the actions of the British officer, the reputed flashpoint for the rebellion was that Indian soldiers found it against their religion to bite off the paper cartridges greased with animal fat, which held the powder for their guns – an act that violated their reverence for animals (Victorian Military Society website). The lion's decline in India was not the result of public pressure against it as a threat to their livestock or themselves.

Cheetah

The Cheetah is an outlier, it being the only native predator to have become extinct within India. The primary cause of its extirpation, though, certainly wasn't the danger it posed to humans. The cheetah was by far the least threatening of India's large mammalian predators, both to humans directly as well as to their property, in this case primarily goats. Indications are that India's cheetahs, which became extinct only recently, in 1947, succumbed primarily as a result of over-hunting by British colonial officers and over-exploitation by Moghul emperors who used them for pleasure-hunting. The Indian Moghul emperor Akbar, for example, in the 17th century, was reported by his son to have kept 9,000 cheetahs over the course of his lifetime, all taken from the wild, to assist in flushing game during royal hunts (Divyabhanusinh 1995). It is important to keep in mind that both the British and the Moghuls occupied India for approximately 100 and 300 years respectively. There are no indications that India's general populace negatively influenced the cheetah's status. Efforts are underway in India today to reestablish the cheetah.

Gray Wolf

Lastly, among India's major predators is the gray wolf – the same species whose population was virtually exterminated in the US's contiguous 48 states in the early 1900s. Its status in India? Numbers are estimated at 2,000-3,000 individuals, the population limited to open arid lands, deserts, and sal forests across northern India (Sharma et al. 2004; Wolf Worlds website). As in the US, this wolf is a threat to livestock – cattle, sheep, and goats. Interestingly, it appears to be a much greater threat to humans in India than in North America. Kishan Singh Rajpurohit documents the killing of 60 young children from 1993 to 1995 in Hazaribagh, Bihar state alone where such attacks are referred to as "child lifting." Matters were much worse a century earlier. In 1878, British officials recorded 624 human killings by wolves (Burns 1996). Despite being decidedly more dangerous in India, the gray wolf was never deliberately exterminated nor required reintroduction efforts.

India's Largest Herbivore

ASIAN ELEPHANT

What about India's largest herbivore, the Asian elephant? For those of us raised in the United States, the Asian elephant is that wonderful, docile creature we see in zoos and circuses which humors us with its performance of masterful tricks, its obedience, and its trunk's awesome utility. In its natural environment, however, the Asian elephant can be a substantial pest. To the regret of many an Indian farmer, rice, corn, sugar cane, and bananas, along with many other subsistence crops, are favorite morsels for the elephant. As such, these animals go out of their way to feed upon them. Nighttime forays by elephants into lands adjacent to their native forests to sample the buffet of crops planted there are regular occurrences during the period of harvest. A local subsistence farmer's entire annual crop may be annihilated in a single night of pachyderm foraging. To thwart the elephants, farmers often build tree platforms from which they overlook their crops as the time for harvest approaches. From these, upon the arrival of elephants, farmers warn one another and collaborate to drive the animals away, though they are far from 100 percent successful in doing so. When an attempt to deter an elephant from its meal goes awry, there may be grave consequences. It is not rare to see rural dwellings that have been virtually demolished by elephants. Beyond that, physical injury or death is scarcely a rare result – elephants are responsible for the deaths of over 400 people per year in India (Report of the elephant taskforce 2010).

Farm plots five, ten, and even 15 or more miles from reserve boundaries are susceptible to these animals. That means tens of millions of subsistence farmers who survive by living off the land face the persistent threat of losing their entire food supply, if not their homes or even their lives, due to elephants. In this respect, the Asian elephant, though perhaps not the direct threat to human life as are the tiger and leopard, is certainly a far greater overall threat to human well-being in India.

Despite all the damage elephants cause, the wild population of elephants was estimated in 2019 to be 28,000 to 31,000 animals (Walk Through India website). This is an all-time low – the result of continued diminution of this large herbivore's habitat and illegal harvesting of its tusks for the international ivory trade. It is due in only the most limited way to retribution for its impositions on humans and their livelihoods. Generally speaking, Indian farmers accept their conflicts with elephants as a challenge to be accepted as part of life. Though killing pest elephants would be easy enough, doing so is not considered an option within the farmer's worldview. Consequently, the elephant in India has faced nothing even remotely like the slaughter of the American bison. In 2010, the Indian government created the Elephant Task Force to address more effectively the threats of human-elephant interactions. The perspective of the task force, with its strong attention to the revered place of the elephant in Indian culture, is worth noting:

- Gajah (the elephant) is a symbol of a search for better compact with nature, our land and our common natural heritage. The rest of Asia and the world look upon us to rise to take a lead. So declaring it the National Heritage Animal will give it due place as emblem of ecological sensitivity. It will also mark recognition for its centrality in our plural cultures, traditions and oral lore.
- Someswara wrote almost eight centuries ago, it is the realm with many elephants in its forests that will be truly most secure.
- India cannot fail Gajah. The latter's survival and ecological security is linked to our very own (Report of the elephant taskforce 2010).

Summary

We have now reviewed the status over time of quite a number of species which serve as important indicators or bellwethers of what has transpired regarding wildlife conservation in the United States and India. From this, there are several important points to note. One which stands out is how much more deadly some of India's animals are to humans and how much greater the threat is, certainly by elephants, to people's livelihoods. India's predators, particularly the tiger and leopard, are dramatically greater threats to humans and their livestock than are their American counterparts. It is one thing to lose one's cow; it is another to lose one's child. Just imagine a predator in the US that required us to take a number of significant precautions to ensure that we were not mauled or killed any time we walked to the barn, the market, or through a forest. Such a predator would not survive long in any community, neighborhood, or environment on American soil.

Second, this inordinately greater threat makes it that much more notable that India's predators and elephants were never subjected to the widespread persecution nor state-organized extermination campaigns their counterparts faced in the United States. And all this despite a dramatically higher human population density, a situation which places large numbers of people into potential conflict with these animals. Clearly, a unique form of accommodation and level of tolerance has occurred.

Third, a primary threat presently faced by most of these predators along with the elephant in India is driven externally by markets abroad, not by domestic decisions or attitudes. Were these external pressures removed, these species would be decidedly more abundant.

Fourth, even with India's enormous human population density, enough habitat has been preserved or is cohabited in an amensal fashion (the wild species not being seriously affected, but the human inhabitants making a sacrifice), to allow for the survival of these remarkable species.

Fifth, India was occupied for approximately 300 years by the Moguls (early 1500s to mid-1800s) and 100 years by the British (mid-1800s to 1947). These two powers were much more abusive to the native wildlife, causing major negative impacts to species such as the tiger and cheetah. Despite four centuries of such rule, a period as lengthy as that from the date the Pilgrims arrived to the present day, the elephant and major predators of India managed to survive much more safely than their US counterparts.

Unique attributes regarding India's wildlife

- Some are extremely dangerous
- Minimal persecution by local entities
- External markets are greatest threat
- Habitat heavily cohabited by humans
- Survive despite centuries of colonization and 5,000 years of civilization

So, what is the point of this comparison? It is to demonstrate that India has proven a better steward of its wildlife than has the United States. In fact, it has demonstrated India to be a far better steward. Many in the US are very proud of the nation's wildlife conservation legacy. Many think it the best in the world. That is their choice – though likely based upon limited perspective. It is for just that reason this comparison is being made and the fact is that, at least when we use predators and major herbivores as a yardstick, indications point clearly to India being a far better steward of its wildlife than the United States.

How Might We Account for India's Success

How might we account for the surprising survival of this Indian wildlife given the much more conflictive conditions there? Probably the primary explanation proffered, one suggested by a renowned associate, is quite simple. India's host of dangerous predators survive because of that nation's level of poverty. The implication here is that India's populace lacks the resources, mobilization skills, or perhaps the drive and resourcefulness to wipe out threatening tigers, leopards, and wolves. That is an interesting thought. Certainly the riveting stories of Jim Corbett, the storied big game hunter of the early 20th century, who stalked and killed man-eating tigers and leopards while "helpless" villagers cowered in their abodes promotes such a sentiment. We might infer that were it not for the steely nerves and raw courage of this British Army colonel, tigers and leopards might have dominated India's masses well into recent times.

The fundamental problem with this argument is that you do not have to search far throughout the historic range of the tiger, leopard, or wolf to discern numerous areas of overlap with impoverished societies where one or more of these species has been exterminated. Iran, Afghanistan, Mongolia, Iraq, Azerbaijan, Turkmenistan, Uzbekistan, Bali – all are examples of countries which were, historically, no more well-off than India, but which have wiped out their tigers. In fact, from one end of the globe to the other, impoverished societies have had no great difficulty killing off predators when they so desired. Poisons are not a recent discovery; knowledge of them dates back many millennia. They also happen to be quite as effective at exterminating animals as they are for killing our fellow man.

Another possible explanation regarding India's notable

conservation successes is that royalty protected this wildlife in their private reserves, thus keeping it from the ravages of India's masses. Certainly royalty helped protect some species, particularly the lion. Contrarily, royalty seemed equally likely to be a greater threat to wildlife than its savior as we saw in the case of the cheetah and has been widely documented for the tiger. A tacit exception was made for kings and nobles to hunt animals, something some did to great excess (Basham 1963).

A third consideration would be not to accept but rather challenge these results by suggesting that this analysis of predators and the largest herbivores is too cursory or narrowly focused to make such sweeping statements regarding wildlife stewardship. Actually, expanding the analysis to a broader array of species will only strengthen the case made here. In Chapter 3 we will touch on this regarding poisonous snakes and in Chapter 4, when we take the Mayflower out for a second voyage, we will see that comparison of the extinct birds of the eastern US and India reflects just as dramatic a story as do the animals discussed thus far.

I believe that the most accurate explanation for India's success in sustaining its wildlife is the values system of its people. It's as simple as that. The Indian populace as a whole, across millennia, has maintained a respect, in fact a reverence for wildlife and nature in general, that has fostered their coexistence and the survival of species which would be found intolerable in many other societies.

Ten Factors of Reputed Importance for Effective Wildlife Conservation

To buttress the case for the above statement, one so central to our discussion and so widely minimized, if not disparaged, within the professional conservation community of the US, it is important that we explore India's successes from an entirely different perspective. This takes us away from the discussion of species as we have undertaken thus far and, instead, focuses on factors which appear to form the foundation for effective conservation.

Factors that form the foundation for effective conservation? We shall bypass the seven tenets of the North American Model of Wildlife Conservation that set off this analysis due to their bias and factors discussed earlier. Instead, it will be productive to look at conservation from a more global perspective, unfettered by the constraints of a United States-based perspective.

To that end, I constructed a simple conservation profile of factors widely perceived as essential for wildlife conservation to be effective. It seems obvious that any country which could rate highly on these priority factors, of which there are ten, should be home to an extraordinarily rich and well conserved national fauna. Contrarily, a country with low ratings would probably have greatly reduced wildlife populations.

Ten factors of reputed importance for wildlife conservation:

1. The larger country's **landmass**, the more space available for various uses, consequently the greater the potential to partition space to sustain wildlife populations.
2. The smaller the size of the **human population** the less likelihood for conflict with wildlife, thus the greater the potential for sustainable wildlife populations.
3. The more **available habitat** for wildlife, the greater the potential for wildlife to survive.
4. The shorter the **period that humans have inhabited the area**, the greater the potential for wildlife to have survived due to a briefer history of contact.
5. The more severe the level of **poverty**, the more likely wildlife will have diminished due to using it for food. Further, wealthier societies can fund wildlife management.
6. The greater its **scientific capacity**, the more likely it is to study, understand and, as a result, conserve its fauna.
7. Stronger **legal frameworks** enhance the rule of law which leads to improved wildlife protection.
8. The greater the capacity to **enforce and implement the law**, the more likely wildlife will be protected and thrive.
9. The more sustainable the **use of natural resources**, the greater the opportunity for wildlife to flourish.
10. The more sensitive the **values, beliefs, and culture** of a society towards wildlife, the greater its survival potential.

Important here is whether scrutiny of these factors could provide insights into what truly are the driving forces for conservation in the US and India. One would think they would help us pinpoint the source of India's success. Actually, they are amazingly telling. We have much to learn from them. For the sake of brevity, I shall focus on those tenets which offer us the greatest insights.

The first two factors – geographic and population sizes – determine a nation's human population density, a factor widely considered highly relevant to wildlife survival. It was mentioned earlier that the US population is presently at 332 million people. As a point of reference, in 1900, it was 76 million. A century earlier, in 1800, it was only 5 million people. The US population is now 60 times the size it was around the year Lewis and Clark set off on their historic expedition in 1804.

India's nearly 1.4 billion people represent one-sixth of the world's population. It now leads the world in this regard. In 1900, India's population was 271 million, thus demonstrating a growth rate similar to that of the US. The 19th century, conversely, was a different story. In 1800, India's population was estimated at 255 million, only slightly less than in 1900. Looking back at US numbers, our population had increased 15-fold over this same period.

The point here is that though India's human population density presently is approximately 10 times that of the United States, if we look back to the era when the US was rapidly expanding westward in the 19th century, and concomitantly decimating its wildlife along the way, India supported as much as 50 times the population of the United States. Cramped into India's smaller geographic area, this accounted for a **population density of 150 times that of the United States!** At this same time, India's wildlife populations were thriving. What do I mean by thriving? As mentioned earlier, it is estimated that **over 100,000 tigers**

survived in India into the early 20th century.

This is not a minor point. These extraordinary facts suggest that the importance of low human population densities as a necessary component for wildlife sustainability are ill-founded. This widely embraced sentiment is a myth.

Factor three appears to be a no-brainer – the more available habitat for wildlife, the greater the potential for wildlife to survive. Based on the above discussion regarding the vast disparity in human population densities between the two countries, there is no question that the US possesses substantially more available wildlife habitat than does India. The possession of more extensive habitat does not seem to matter.

With respect to **factors four, five, and six,** the US is dramatically better placed regarding their achievement. India has been inhabited in a highly civilized state for approximately 5,000 years (Bhasam 1954), thus coping for millennia with wildlife conflict. The US has coped for scarcely a few centuries – tenet four. The number of impoverished people in India in recent times has numbered in the many hundreds of millions, far more than the entire population of the United States, thus the pressure on wildlife for food in India should have long been overwhelming – tenet five. While scientific expertise is expanding rapidly in India, the US remains far advanced in this capacity thus benefiting from greater knowledge and management techniques to sustain wildlife – tenet six. Our earlier finding that India's wildlife is so much better conserved than that of the US suggests that these "truths," formerly considered essential prerequisites for conservation to be achieved are, in fact, myths.

Factor seven and eight are related, the former referring to the importance of possessing a strong legal framework of effective laws coupled with the latter, the capacity to enforce and implement such laws. There would be little argument that the US is better endowed regarding these two tenets.

Factor nine is complicated. It pertains to the sustainable use of resources – both living, renewable resources such as wildlife, and non-renewable ones.

The term "ecological footprint" is widely used to reflect the impact of an individual or a nation on the resources of our planet. The term incorporates the natural resources a population consumes along with its ability to absorb the waste it creates. The US ranks fifth among countries with a population of one million or more. The footprint of the average US citizen is about 50 percent larger than her/his counterpart in most European countries. The US has more suburban sprawl, less public transportation, and uses more energy and water per person than other comparable developed countries (World Population Review website). Americans are using twice the renewable natural resources and services than can be regenerated within the country's borders.

In contrast, India places in the bottom 25 countries in this measure. (World Population Review website). Its per capita impact on natural resources consumption is very small.

Where does India stand regarding this tenet of sustainable natural resource use, specifically with regard to its wildlife? I believe we have already had a good hint of this from our previous analysis of India's predators and the Asian elephant. More examples shall be presented in the following chapter. As far as the direct treatment of animals is concerned, India treats them, and sustains them, far better than does the United States.

Regarding indirect impacts on wildlife sustainability, India deserves lower marks. The impacts of alien cultures, industrialization, bribery, and unfettered capitalism, coupled with an immense populace, have led to filthy rivers and, in major cities, unbreathable air (Agarwal 2000; Dwivedi 2000; Arunji Gandhi personal comment). It is these conditions that foster the suggestion by Dwivedi and Tiwari (1987) that India is poor at managing its resources. And, in fact, the government may be. But it is not the government alone that interfaces with wildlife. It is the populace that does so, and there is no question that India's populace is far more wildlife-friendly than that of the United States – this despite what birdseed sales may indicate.

Clearly, the issue of sustainable natural resource use is central to effective conservation practice and a more detailed analysis likely could offer additional insights. Nevertheless, we have addressed the matter to the extent necessary for this discussion. I believe it is clear that India, not the US, ranks higher regarding this factor – unquestionably from an overall resources perspective, but also from that of specific wildlife sustainability as well.

We have now identified a single tenet, the sustainable use of resources, which India implements more effectively than does the United States. The question then is whether this tenet provides a more profound explanation and understanding of India's success at sustaining its wildlife.

It helps, decidedly. As we have discussed, human consumption of resources, particularly their unsustainable consumption, does not occur without a concomitant environmental impact. By living more sustainably overall, the impacts of India's population and economic growth upon its wildlife resources are clearly lessened. How much so? That is difficult to say. Can it account for the dramatic contrast we see between the faunas of the US and that nation? Not at all. The survival of tigers, leopards and elephants, extraordinarily dangerous animals, in the midst of such a densely populated nation as India simply cannot be explained by the practice of sustainable resource use. There is clearly more to this picture.

It is in the final factor, tenet number ten, that I believe we find the key which unlocks the door to understanding what, up until now, has appeared a paradox concerning India's prodigious wildlife conservation achievements.

Factor ten pertains to a society's values and beliefs. What is that society's culture, norms, and mores with regard to wildlife? What is wildlife's standing with regard to the manner in which the society views the world? Where does wildlife fit in people's hearts and minds? How should wildlife be treated? What is its purpose? These often deep-seated feelings, feelings passed on through our culture, our families, our religion, our education, are what factor ten is about and, in fact, what appears the best explanation for India's impressive wildlife stewardship – its values system.

Indian culture, due to its high sensitivity towards living resources, has enabled the survival of predators and herbivores that would not possibly have survived had Western values dominated the Indian subcontinent.

Where does this leave us? Actually, it leaves us in a critical and powerful place. Accepting that societal values are more essential to achieving conservation than are all other factors combined, enables us to lay out a much more effectively focused and impactful approach to conservation with the potential to be a game-changer in this field. The other option, to deny this reality, leaves us to continue on with the fundamentally flawed practices widely underway all around us today.

To reinforce this conclusion the next chapter will delve into how the US and India value wildlife. It is worthwhile that we do this so as to better appreciate the distinctiveness of India's culture which helps us to reaffirm that societal values, and the behaviors built upon them, are not merely an important element of conservation, they are its essence.

CHAPTER 3

VALUES AND BELIEFS: THE DIFFERENCE-MAKER

This chapter provides some additional background on the development of wildlife conservation in the US. We then look at India with regard to some present-day practices as they relate to wildlife and plants. A brief comparison to China is included because historically that country supported many of the same species as India, but today the status of wildlife in the two countries is remarkably different. Why? All of this will reinforce the basic point that societal values are a game-changer when it comes to achieving conservation.

Conservation Values in the US

What have been the dominant values and beliefs in the United States when it comes to conservation, particularly of wildlife? Historically, our nation has long perceived wildlife as being there for our use. This derived both from the Christian ethic of its colonizers which characterized the living things around us as having been created for our use, as well as from the remarkable abundance of wildlife found here, giving the impression that it could be exploited without impact. Perhaps the most critical element identified by the North American Model for Wildlife Conservation was putting constraints on this use via the termination of market hunting.

By the late 19th century most of the country's wildlife that had become endangered – American bison, passenger pigeon, Carolina parakeet, Eskimo curlew and many other shorebirds, various ducks and geese – had gotten to their precarious state as a result of tremendous hunting pressure to fill the demand in markets for their meat. When European colonists first set foot on

North American shores, it was the bounty of the region's natural resources, including its wildlife, which was essential to sustaining them and to enabling their early growth and expansion. New York City, as an example, began as an outpost of the Dutch West India Company to trap and procure furs, primarily beaver, for export back to Europe. The Atlantic cod was harvested intensively by the Vikings prior to Columbus' arrival in the New World and by the Basques prior to the "discovery" of the St. Lawrence River by Jacques Cartier in 1534. The fish were salted, cured, and shipped back to Europe as a major item for trade, a trade which later was to become an important component of the slave trade triangle. Cod trade from the US to the Caribbean was so significant that ackee and saltfish (cod) became the national dish of Jamaica, a country located over 1,000 miles outside the range of that fish.

Though in the US terrestrial wildlife is still harvested today, taking it for commercial meat markets is virtually non-existent. This meat now comes from domesticated animals. Interestingly, such is not true in the case of aquatic organisms – specifically fish. As a result, many popular food fish in the US suffered severe declines. To address this crisis, Congress amended in 1996 and 2006 the Magnuson-Stevens Fishery Conservation and Management Act of 1976 to require that overfished fisheries be rebuilt to healthy levels. Implementation of this Act has resulted in recovery of some stocks (Sewell 2013).

The harvesting of terrestrial wildlife for commercial purposes other than food – alligators for their skins, beaver, mink, muskrat, and other mammals for their pelts, and so on – has, to some extent, been sustainable, primarily thanks to governmental regulation and shifts in consumer preferences away from the skins of these animals.

It was through mobilization of the US hunting community over a century ago that major steps were taken to establish legal controls on hunting so as to conserve wildlife. The string of conservation accomplishments in this regard is impressive. Among these was passage of the Migratory Bird Treaty Act of 1918 which authorized the federal government to set hunting regulations; creation of the Federal Duck Stamp in 1934 requiring purchase of the stamp by all waterfowl hunters, the proceeds go towards the conservation of duck habitats; and passage of the Pittman-Robertson Wildlife Restoration Act of 1937 which taxes purchases of hunting equipment, the revenues of which are distributed among the states to manage game animals.

The active involvement of the hunting community, however, led to a conservation bias toward game rather than non-game species, a focus on animals which are hunted. It was the game animals towards which nearly all conservation efforts were dedicated for the better part of the 20th century. For all intents and purposes, it was not until passage of the Endangered Species Act in 1973 that major attention was turned towards a few non-game species.

One consequence of the focus on game animals was that a substantial number of species of no import to hunters were nearing the brink of extinction. I recall well from my years as a field biologist in Puerto Rico, leading up to passage of the Endangered Species Act, of prodigious efforts being made to conserve migratory waterfowl while endemic species on the verge of sinking into oblivion – the yellow-shouldered blackbird, Puerto Rican toad, Puerto Rican nightjar – to name a few, received virtually no attention. Worse, at least one magnificent mangrove swamp, important as a fish nursery and for many other valuable ecological reasons, was drained in a desperate, and mostly unsuccessful attempt to create waterfowl habitat for hunting. Other examples in the literature and the US Endangered Species Act testify to the serious impact of this narrow focus on game species – 78 mammals, 105 birds, 48 reptiles, 36 amphibians, 139 fish, 312 invertebrates, and 942 plants are listed in the ESA as either endangered or threatened at present (US Fish and Wildlife Service website).

A second consequence, evident up until passage of the Endangered Species Act, and still persisting today, was that any species believed to prey upon or be a threat to game animals was treated as a pest, or worse, and was deliberately persecuted. This included not only top predators such as the mountain lion and wolf which prey upon elk and deer, but also relatively innocuous species such as the raccoon, fox, skunk, and weasel which were perceived to threaten duck reproduction. Beyond that, all predators, competitors for resources, or species otherwise perceived as a threat to a rancher's cows or a farmer's chickens were often treated as vermin. In addition to a number of the above species were included prairie dogs, great-horned owls, and snakes, among others. This negative attitude towards many species was reflected in the Animal Damage Control Act of 1931, referred to in the previous chapter. Despite the Endangered Species Act, persecution of common species considered pests under the Animal Damage Control Act – coyotes, prairie dogs, foxes, raccoons, and the like – continues unabated to this day.

"Change is the law of life. And those who look only to the past or present are certain to miss the future."

– JOHN F. KENNEDY

A prime example of the negativity with which some Americans view wildlife is the Texas Sweetwater Jaycees Rattlesnake Round-up, one of many such "festivals," in which thousands of rattlesnakes are slaughtered in a single event. The event proponents ingeniously, and ingenuously, have the gall to suggest that such events help sustain the balance of nature (Weir 1992). These killing contests, legal in 43 states, are also held to kill bobcats, coyotes, foxes, prairie dogs, rabbits, raccoons, and squirrels among other species. Within but a few years following the official 2011 recovery designation of the gray wolf in parts of the US, over 2,500 wolves were killed either for fun or out of resentment, those numbers increasing regularly (Hance 2014).

All this is to say that historically in the US the "utilitarian" perspective towards wildlife was, and to a large extent remains, dominant. In this view, wildlife is there for our use and benefit, and what is not usable or useful to us is ignored – or worse. I placed "utilitarian" in quotation marks because it is likely that few or none of the animals slaughtered in these killing contests are "utilized." Having participated in a public hearing regarding a coyote killing contest in Massachusetts, it was evident that many in favor of slaughtering these animals were doing so for fun. One speaker referred to killing coyotes as being just as justified as any other contest, and then went on to compare it to a beauty pageant! As to the Fish and Game Department's response? They explained that attitudes were not their concern. Their concern was the biology of coyote populations – would such contests negatively impact coyote numbers. You'll notice that this total dedication to species biology by the Fish and Game Department was identical to the one I mentioned earlier by the US Fish and Wildlife Service director to his Mexican counterpart. This was not coincidental.

Is this a fair picture of the historically prevalent view in the US toward the value of wildlife? I believe it is quite close. Virtually since its founding, history demonstrates that the dominant values system in the US as it relates to wildlife has been predominantly utilitarian. If wildlife is useful for commerce or sport – let's manage and conserve it. If it's a nuisance or a pest – let's explore options to get rid of it. Of course, this view is not representative of the entire populace. The values expressed by Muir and Leopold remain as a small flame in the mindsets of some. The first Earth Day, held in 1970, was a major pulse of environmental activism. Bird watching, backyard bird feeding, and nature photography are growing – and growing rapidly. Meanwhile, the mindset in many parts of the country, particularly of professionals in the state agencies and in hunting-related organizations, has scarcely changed.

"The world hates change, yet it is the only thing that has brought progress." – CHARLES KETTERING

That said, attitudes among professionals are gradually shifting over time towards a more holistic perspective regarding our nation's wildlife. For example, the nation's wildlife refuge system, once primarily focused on duck production, in 2005 revised its goals to conserve a broader spectrum of the nation's living resources. Also, in recent decades one can see, state by state, a noticeable expansion of non-game programs. But this shift, though positive, has been relatively small – far too inadequate to address today's conservation needs. We still have a long way to go.

Conservation Values in India

How about India? How do perspectives and values concerning wildlife differ there? I am not an expert on Indian ethics or culture, but my personal experiences and interactions are telling and supported strongly in the literature.

A few years back I was conversing with an Indian woman raised in one of that nation's major cities. She had scarcely been exposed to nature as a child nor did she have any particular inclination towards wild things. Was there anything about her culture while growing up in India that she found different from her experience in the US, I asked. Her response: "My mother taught me not to touch plants at night. After all, they have to sleep too." Now, isn't that an incredibly sensitive perspective on nature fostered in a densely populated Indian city?

In subsequent discussions with Indian colleagues, I have inquired about this perspective concerning plants. Nearly unanimously their response has been: "Why of course! Who would do a thing like that?" It was as though everyone knows the impropriety of touching a plant after dark. Clearly, Indians show a sensitivity and respect for plants beyond that practiced in the US or the Western World.

KOLAM

Primarily in southern India, many women prepare a Kolam each day in the belief that it will bring prosperity to the home. Other names for this adornment are Muggu, Rangoli, and Tarai Alangaram. Its creation involves using a colored powder to make a drawing on the ground or path at the home's entranceway. Traditionally, white, or sometimes colored, ground rice is used with the intent that ants will come and dine on the offering and not have to work so hard for a meal. The rice powder is also intended to attract birds and other living things to one's home as a sign of harmonious coexistence (Aggarwal 2000; Nagarajan 2000). This tradition is an extension of the Indian ethos of oneness with nature fostering the well-being of all living entities – recognizing the whole world as one single family (Sarma 2019). This tradition dates back 5,000 years. How many westerners would find it appealing to attract ants to our houses?

Most westerners are familiar with respect for cows in India. In Indian culture, the cow is regarded as a symbol of human responsibility to non-human life and reflects reverence to those forms. Blessings sought daily for cows leads to happiness and peace at all levels of human existence (Seshagiri Rao 2000). Feeding a cow, therefore, is an act of worship. Eating one, or meat of any animal, is a sin. Other animals, too, are worshiped, many having sacred stature.

While travelling in the state of Karnataka in southern India, I learned that elephants can sense water at great distances, perhaps ten miles or more, and that they will stray at night from the forests in which they reside during the day to feed on crops of villagers eking out a living adjacent to the reserves. In fact, between 1980 and 2003, 1,150 people were reported killed by elephants just in Northeast India (Choudhury 2004). How can they be tolerated? As one destitute subsistence farmer, who had recently lost his entire crop to elephants, put it, "Yes, I lost my entire harvest, but how can I deny an elephant its right to eat?"

Once, while deep in the forest of a reserve located in the Western Ghats, a mountain chain in southwestern India, I encountered a procession of people who had appeared out of nowhere. The group, containing 70 or more men, women, and children, evidently had some clear destination in mind, though I could not fathom what it might be at the time. It was also evident that most group members were not well off, were not prepared for hiking in the forest, and were suffering as a result. Upon inquiring, I learned that the group was on its way to pay homage to a tree – a demonstration of the respect Indians have for nature in a broader sense, not just wildlife, and that they will sacrifice for it. Sacred texts value them so highly that their destruction is connected with doomsday scenarios (Narayanan 2008).

Learning the purpose of this procession, out of the ordinary from a westerner's perspective, reminded me of a case I had heard of which, to my knowledge, has no parallel in Western civilization. Back in 1753, near Jodhpur in northwestern India, the woman Amrita Devi is said to have led members of the Bishnoi sect to protect the *khejri* trees of their village from destruction by the soldiers of the king. Three hundred and sixty-three Bishnoi gave their lives in this effort (Rukmani 2000). This included 36 married couples, one of which was a pair of newlyweds.

This example is not reflective of an environmentally sensitive indigenous people being conquered by more powerful invaders. The Bishnoi were and remain an integral part of Indian society. They were defending a traditional component of their worldview which includes reverence for trees (Rukmani 2000) and the concept that forests support plants of medicinal and economic value and are where one can find enlightenment (Lee 2000). Likely the group I observed was on a trek related to this vision.

What jumps out from these examples? Most Indians generally have a deep respect for wildlife and nature. This respect is not confined solely to the big, warm, and fuzzy variety. It extends as much to ants and plants, as it does to cows and elephants. There is nothing in this sentiment that distinguishes utilitarian versus non-utilitarian wildlife – a game species versus a pest. In India, all wildlife is respected and worthy of appreciation.

It goes without saying that not all Indians believe so strongly about the environment in this way. Indian maharajas famously used their minions to drive tigers in their direction so that they might shoot or spear them. Apparently, a tacit exception was made for nobles (Bhasam 1954). An indication of India's diversity is that over 800 native languages are spoken in India, with 22 of them designated as "official" by India's constitution. With them come diverse beliefs and practices.

The examples presented above are indicative of the compassion towards wildlife and plants embedded in Indian values and beliefs. Many Indians do not not accept the killing of animals or the eating of meat. Many Indian scriptures emphasize that harming or killing other creatures will deny one the grace of god (Dwivedi 2000). Resultantly, many Indians are vegetarian. This precept influences various elements of Indian life far beyond tolerance of any personal discomfort or inconvenience wildlife might cause. It touches upon things as small as the right and wrong way to display an elephant statue in one's home, to as substantive as basic dietary practices. On the subject of wildlife, Mahatma Gandhi once said, "The greatness of a nation and its moral progress can be judged by the way its animals are treated."

Perhaps more dramatic than any of these examples is the status of snakes in India. I referred earlier to snake killing contests in the US, a widespread practice. After all, about 7-8,000 Americans are bitten by poisonous snakes annually, approximately five of whom die. How does this compare to India? India has a diverse array of poison snakes including various cobras and vipers. In total their bites kill over 50,000 Indians per year. That is a remarkable jump from the 5 US deaths. What is India's response? India celebrates snakes. In some areas killing them is a crime and a penance of up to a year may be served. The most dangerous snake – the spectacled cobra – is the most celebrated! The cobra is worshipped over much of India and festivals honoring it are held each year. Does this mean that no snakes are killed in India? Of course not. But the overall cultural perspective on snakes is one of respect, not scorn.

Fundamentally, in the case of India, its unique culture apparently has created a foundation of values and norms deferential to, if not revering of, wildlife.

India is no environmental paradise. A few short years ago, though to a lesser extent today, India's cities were often shrouded in some of the world's worst smog, rivers and other waterways are filthy. This was discussed briefly earlier, and the likely causes mentioned. In addition to those, Indian scholars offer explanations for this seemingly paradoxical circumstance to the belief that rivers such as the Ganges are so powerful that it is impossible to pollute them, because they heal themselves despite evidence to the contrary (Alley 2008; Narayanan 2008). Regardless, what is important is to draw lessons concerning conservation from what has worked and, with respect to wildlife, the India case is indicative of the power of underlying values in determining society's practices. Were it not for such values, India might long ago have become denuded of wildlife. India serves as a prime example in support of author Gene Anderson's point in Ecologies of the Heart, that the effective management of resources necessitates "a direct, emotional religiously socialized tie to the resources in question" (Orr 2002).

Conservation Values in China

As a case in point, we only have to look at neighboring China, a country which housed all the same species as India. The contrast is striking. China has fewer than 300 elephants; tiger estimates suggest no more than 50 animals survive in China's wild. The leopard, a species that, like the others, is known to have formerly been widespread in China, now numbers in the order of 100 animals. As for the gray wolf, it is believed to still survive in China in small numbers (Wang et al. 2016).

China has three primary, long-standing religions. The most widely accepted of these, Confucianism, is relatively human-centered, while the other two, Taoism and Buddhism, center on a connectedness and respect for nature (Tucker 2008). Interestingly, when China does want to promote some environmental goal, the government's Ministry of Environmental Protection appeals to that nation's spiritual heritage which, in the past, was much more environmentally sensitive (Gardner 2010).

This is no hollow gesture. Strongly supportive of the case we have made for India, namely, that societal values are the most powerful of tools, China's history too, serves as an example. Despite all the China-bashing presently taking place due to some of that nation's recent practices, particularly elephant ivory consumption, as recently as the 1950s, the South China tiger, now nearly extinct in the wild, numbered approximately 4,000 individuals. At that time, the tiger, along with other predators, was declared by the new Maoist government to be an "enemy of the people" for its threat to livestock and villagers. The heavy persecution which followed resulted in tiger numbers dropping to a mere 200 animals by 1982. Though the Chinese government reversed its position and banned tiger hunting in 1977, this apparently was too late to save the relic population.

Tragically, not only has China nearly wiped out its populations of the Asian species we have discussed, it has done the same for many other species on which its large population has historically depended for food and other uses such as traditional medicine. Increasingly, China has sought to import wildlife and their products to make up for the unsustainable harvesting and subsequent demise of its own wildlife heritage. This trade has negatively affected wildlife populations around the globe, India's among them. Ironically, what has an indignant global conservation community blamed for this perceived abuse?

China's cultural values. Westerners are quick to recognize the downside of values, yet reticent to accept the positive contributions they can make.

Societal Values – The Game Changer

Thus, we see the power of cultural values. They work both ways – for better or for worse – but either way, their impact is nothing less than overwhelming. Nevertheless, the professional conservation community with which I have worked – federal and state agencies along with many international non-governmental organizations in the US and abroad – has typically failed to recognize this point and has not given cultural values their due. Such is especially the case in the US and the developed world where much of this community generally believes that biological and ecological facts drive people's decision-making, not the feelings in their guts and their hearts. This is a serious flaw. Until it is remedied, conservation efforts will continue to achieve limited success.

Within the professional conservation community, our bias against cultural values is so strong that on the one hand we blame China's cultural values for their negative impacts, both in that country and abroad, while, on the other hand – India's – we choose to be blind to the benefits resultant from cultural values which have sustained a wildlife heritage of unique importance to the entire world. For those of you who may believe I am overstating this point, I urge you to review the augustly titled work, The Development of International Principles and Practices of Wildlife Research and Management: Asian and American Approaches by Berwick and Saharia (1995). This 481-page tome, despite containing two chapters on the human dimensions of wildlife management, fails to mention at any point the positive nature of India's culture as it relates to wildlife. Fundamentally, the work, though written by a number of well-known US and Indian wildlife managers, is basically a treatise on the application of US wildlife management techniques in India. As far as conservation is concerned, this book is like driving down a one-way street – in the wrong direction!

Dwivedi and Tiwari (1987), two Indian scholars, considered India's environmental record extremely poor based on pollution levels, thus leading others in the field such as Gardner and Stern (2002) to misinterpret the influence of religious values on societal behavior. This has led Richard Foltz (2008) to comment, "We may even find that the solutions to our problems are already available – solutions that have existed for centuries, but the cultural blinders imposed by the dominant ideology have prevented us from seeing them."

The importance of values is not unique to wildlife conservation. It even is acknowledged in the corporate world. As IBM's former CEO, Louis Gerstner (2002), put it, "culture isn't just one aspect of the game – it is the game."

The ultimate importance of the present discussion is to recognize the power of people's ethics and values – for our purposes, those relating to wildlife and conservation more generally. It corroborates the insight of Christopher Flavin (2010), president emeritus of the Worldwatch Institute, that "one dimension of our environmental dilemma remains largely neglected: its cultural roots." This has been stated even more powerfully by the renowned zoologist George Schaller (2011): "Conservation is based on moral values, not scientific ones, on beauty, ethics, and religion, without which it cannot sustain itself." The wildlife management profession in a strict sense, meaning state and federal wildlife professionals, recognized the "human" component of their work to some degree through the formation of a "human dimensions" component in the 1970s, a journal in 1996, and a fine textbook in 2001 devoted to that theme. Unfortunately, this effort has grown slowly, is heavily research based, has been driven largely as a problem-avoidance mechanism, and is quite marginalized within the profession.

Returning to the 10 conservation elements analyzed earlier, we found the US to be more favorably disposed regarding eight of them – India on but two. Were each of these tenets of equal importance, the US would harbor a dramatically more intact and well conserved fauna. We found virtually the opposite to be the case. This points to a society's values, ethics, and norms being an inordinately important attribute affecting the status of wildlife in a country or region. Arguably, it is a factor which overrides all others. It is fair to contend, if our earlier review is accurate, that a society's values have more of an impact on conservation **than all other factors combined.**

This is no trivial conclusion. The implications of this view turn the United States' domestic approaches to conservation, as well as nearly all international ones, on their ear. Recognition of the importance of societal values, typically an afterthought in much conservation practice today, requires that we completely rethink how conservation is achieved. Addressing conservation values will have to come first in our thinking as well as receive dramatically increased attention. As environmental sociologist Thomas Heberlein (2012) succinctly stated in his book, *Navigating Environmental Attitudes*, "values are the basis for many attitudes and play a major role in discussing pro-environmental behavior, as they should."

Such a conclusion may be difficult for many of those active in conservation to accept. After all, most of us have seen and experienced some level of success, even if relatively limited in the face of the stream of emerging crises facing us. In the West, environmental professionals have been trained to believe that simply citing biological and ecological data repeatedly and extensively is what is needed to effect change in people's conservation-related practices. Besides, the thinking goes, since all of our colleagues support what we are doing, it is presumed we must be doing something right. Still, when a nation like India has invested financially so much less in sustaining its wildlife than the US, yet has achieved such superior results, it gives one pause. The US Fish and Wildlife Service alone spends over $1.5 billion annually in its conservation efforts. The National Park Service over $2.5 billion. The Forest Service nearly $5 billion. The state fish and game agencies at least $1 billion more. And this does not count the contributions of the myriad non-governmental organizations too numerous to mention. Likely, the level of wildlife conservation expenditures made in India over the past century would not amount to a single year's worth in the United States. If we do not recognize the critical importance of values, conservation efforts in the US will never achieve the success for which we strive. Environmental messages, studies indicate, often fail to produce behavior change due to their lack of alignment with the values of most of our populace (Schultz and Zelezny 2003; Saunders et al. 2006).

Summary

The discussion in this chapter, though relatively brief, is intended to emphasize a basic conservation message – an absolutely fundamental one: **Whether a small community or a large nation, its values system is by far the single most important factor in determining its conservation destiny. Money is no substitute. Biological data is no substitute. There is no substitute. We ignore this precept at our peril.**

Does this mean existing conservation efforts should be abandoned? No. There are worthwhile conservation initiatives underway all over the United States and the world. What it does mean though is that existing efforts should be reassessed with a focus on conservation values at the forefront of discussions. It means, in addition, that we should launch comprehensive values-focused conservation initiatives immediately. Just how we might achieve some of these objectives will be the subject of much of the remainder of this book.

Is this asking a lot? Certainly it is. Can it be done? I believe so. Frankly, we have no choice if we are to have a fighting chance to conserve our wildlife for future generations to enjoy.

Before delving into more effective conservation strategies, let's recolonize the US with a different culture.

CHAPTER 4

REVOYAGE OF THE MAYFLOWER

As mentioned in the Introduction, it is difficult to break down long-standing myths. This is especially the case when they are self-serving. Given that such is the case before us, let's look at this issue of the importance of cultural values from several perspectives.

A particularly visual frame of reference derives from asking the questions, "What if the Pilgrims had not been on the *Mayflower?* What if the first colonists to arrive on the shores of the New World had been from India? What would wildlife conservation look like in the United States under this twist of fate?"

It is widely accepted that the Pilgrims emigrated to the New World in the face of religious persecution. They signed the famous *Mayflower Compact* shortly upon arrival through which they agreed to collaborate for the good of the colony and be obedient to its leadership. Being highly religious Christians, the *Bible* served as their primary moral compass and book of learning. The *Bible* is clear that only humans possess a soul. It also contains language (Genesis 1:26) to the effect that man should "have dominion over the fish of the sea, over the birds of the air, and over the cattle, over all the earth and over every creeping thing that creeps on the earth." Despite more nature-tolerant language in other passages, it seems evident that dominion over the earth is the heritage passed on from generation to generation in the US since the Pilgrims first landed.

Along a very different vein, a basic Indian belief is that *all* living things have a soul and that god resides in that soul thus, no species has the right to interfere with the rights of another. Another fundamental Indian belief is that of rebirth – whereby a person is reborn as an animal, bird, or insect. This requires that other species be given not only respect, but reverence. To that end, many Indian nationals are vegetarians.

With these widely differing perceptions as a frame of reference, a telling exercise would be to examine those animal species (all of which happen to be birds) that have become extinct in the eastern US since the fateful day in 1620 when Plymouth Rock was first set foot upon by the most famous of this nation's early colonists – the Pilgrims.

Extinct birds of the United States

Heath Hen

To that end, why not begin with a bird reputedly served at this nation's first Thanksgiving? No, I am not referring to the wild turkey, but rather to the partridges that were also said to be part of the feast. The term "partridge" is a generic one and could pertain to a number of bird species, but in this case almost certainly referred to the heath hen, a fairly large chicken-like bird common during those times in scrubby barrens from coastal New England south to northern Virginia. A subspecies of the greater prairie-chicken, or believed by some to be a distinct species of its own, this bird was so common around Plymouth that it later became a staple in the diet of servants. The bird was so heavily hunted for food into the 19th century that, coupled with the impacts of habitat loss, it was extirpated from the mainland by 1870. The only surviving population, on Martha's Vineyard off Cape Cod, managed to survive until 1932 when it too succumbed, resulting in the extinction of the bird.

Would such have been the bird's fate had emigrants from India landed on Plymouth Rock? Not likely. I can find no instances of widespread slaughtering of animals for food within Indian civilization.

Consequently had emigrants from India voyaged on the *Mayflower*, it is likely no turkey or partridge would have been served at the first Thanksgiving, or any thereafter. The heath hen would have not been hunted at all, and it is fair to say would have had an excellent chance to still be with us today. Doubtless, its numbers would have seriously declined due to destruction of its habitat combined with impacts from introduced animals associated with humans, such as feral cats and stray dogs. Such has been the fate of the northern bobwhite, a similar-type ground-nesting bird that has undergone a serious decline. But, being threatened is not at all the same as being extinct. More than likely, had emigrants from India transited on the *Mayflower*, not Pilgrims, the heath hen would still inhabit small parcels of habitat in the northeastern United States.

Passenger Pigeon

A bird not reportedly served at the first Thanksgiving was the passenger pigeon, a bird that likely migrated south from Plymouth during the fall of each year. Believed to have numbered in the billions, it is widely acknowledged as having been the most common bird in North America in the 1600s. The passenger pigeon was found across North America east of the Rocky Mountains wherever deciduous forests occurred. Unbelievably, the bird is now extinct. Why so? Commercial hunting was a major, if not the most important, factor. Expansion of the railroads played a significant role in enabling the transport of birds to urban markets. For example, Plattsburg, New York, in 1851 alone, is reported to have shipped approximately 1.8 million passenger pigeons to larger cities. A single hunter reportedly killed three million birds during his career. As if all of this were not bad enough, the birds were used for pig fodder and as targets during shooting tournaments. Thirty-thousand were killed in a single competition. By 1901, the once ubiquitous passenger pigeon was extinct in the wild. Martha, the last surviving individual, succumbed in the Cincinnati Zoo in 1914.

It is difficult to doubt that had Eastern Hemisphere Indian rather than Puritan values arrived to North America's shores aboard the Mayflower, passenger pigeons would still abound throughout our eastern forests. Might some other catastrophe have befallen the bird that drove it to extinction? We can speculate, but there is no reason to believe such a scenario to be likely. More plausible is that the passenger pigeon would have remained abundant and as familiar to US citizens as the introduced pigeon that widely inhabits our city streets.

Great Auk

One of the most widely known North American extinctions since the arrival of the Pilgrims is that of the great auk. It was also the first. The closest thing to a Northern Hemisphere version of a penguin, this large, black-and-white, flightless seabird inhabited frigid coastal waters of the North Atlantic coast, ranging as far south in winter as none other than the vicinity of Plymouth. Did the Pilgrims feast on this bird? Not to my knowledge. But, that did not keep their descendants farther north from doing so. Killing of great auks did not stop with harvesting them for food. Often they were used for bait. Their soft down feathers were excellent stuffing for pillows and quilts. Unfortunately, instead of harvesting the down sustainably, the feathers were ripped from the bird's breasts leaving them to die from exposure. Aaron Thomas of the MHS *Boston* in 1794 describes the defeathering, as well as the birds being burned alive for fuel:

"If you come for their feathers you do not give yourself the trouble of killing them, but lay hold of one and pluck the best of the feathers. You then turn the poor Penguin adrift, with his skin half naked and torn off, to perish at his leisure. This is not a very humane method but it is the common practice. While you abide on this island you are in the constant practice of horrid cruelties for you not only skin them alive, but you burn them alive also to cook their bodies with. You take a kettle with you into which you put a Penguin or two, you kindle a fire under it, and this fire is absolutely made of the unfortunate Penguins themselves. Their bodies being oily soon produce a flame; there is no wood on the island."

It is reported that the last great auk was killed in 1844 by three fishermen. After catching the bird, they bound it to their boat, stoned, and then crushed it, apparently for some superstitious reason.

As with the passenger pigeon, such a scenario would have been decidedly unlikely had Indian nationals been aboard the *Mayflower.* The accidental introduction of rats to the North Atlantic nesting islands of the great auk would have reduced its numbers considerably from the million or more once believed to occur. But, extinction? I do not believe so. Had emigrants from India arrived on the *Mayflower,* birdwatchers today from many parts of the country would be trekking in the dead of winter to the outer beaches of Cape Cod, and promontories further north, in the hopes of spying this most spectacular of North Atlantic seabirds plying the region's rough seas as if on a calm lake.

Eskimo Curlew

As the Pilgrims began harvesting their crops in the early fall, it would not have been rare to encounter, from time to time, flocks of an unusual bird with a long, noticeably curved beak, feeding on grasshoppers and other insects infesting their fields and other open areas around Plymouth. Of course, the Pilgrims would have no way of knowing that this bird, the Eskimo curlew, had just flown thousands of miles from its breeding grounds far north on the tundra of the Alaskan and Canadian Arctic and was only pausing prior to flying many more thousands of miles, reputedly non-stop – much of it over open ocean – to the pampas of southern South America. Centuries were to pass before the Eskimo curlew's migration was to be recognized as one of the most spectacular in the world of birds.

It did not take long, in all likelihood, for the Pilgrims to learn that this curlew was good eating. Though scarce along most of the eastern seaboard, with the exception of Labrador and Newfoundland where this migrant staged, the Eskimo curlew was abundant in the Great Plains through which it passed when returning to its breeding grounds in the spring. Here, over time, the curlew became much sought after to a point where, by the late 1800s, it is reported that approximately two million birds were killed per year. The bird's lack of fear, coupled with its habit of travelling in large flocks, made it an easy target.

Not surprisingly, such intense hunting pressure was unsustainable and, despite protective legislation being passed in the early 1900s, the species' decline continued to the point where after 1987, it was to be seen no more. The extinction of the Eskimo curlew was in close association with that of the Rocky Mountain locust which, once ubiquitous in the Great Plains, remarkably became extinct by 1902. The relentless transformation of the plains to agriculture significantly affected both species. Nevertheless, we would be hard pressed not to accept the relentless hunting pressure on the Eskimo curlew as a major, if not the primary, cause of its extinction.

Would emigrants from India have laid the ethical groundwork for such a slaughter, one that is readily traced to the Puritans? One would expect not. Might the Eskimo curlew have become extinct regardless of the hunting pressure? Perhaps, but that is a big ***IF.*** More likely than not, had emigrants from India colonized this continent, Eskimo curlews would have survived at least in sufficient numbers into the 20th century that subsequent conservation efforts would have saved them, just as was done with a number of other plains-dependent animals such as the black-footed ferret, whooping crane, and two species of prairie chickens, among others.

Carolina Parakeet

To this point, we have assessed all the bird species that became extinct which, at the time of the Pilgrims, occurred at least from time to time in the vicinity of Plymouth, Massachusetts. But, the Pilgrims did not confine themselves to Plymouth for long. Rapidly, they and their descendants spread in every direction, ultimately serving as progenitors for settlers far and wide. In all probability, it would not have taken long following arrival of the *Mayflower* before exploration extended far enough west into what is now New York state to encounter the last of the extinct species germane to our discussion. That bird is the Carolina parakeet. Few Americans today probably have any idea that much of the eastern portion of the country was home to a beautifully plumaged parakeet with red face, yellow head, and green body. Despite its beauty though, this parakeet was, like many parrot species, a pest to agriculture as colonists spread throughout the bird's range. No problem. Following upon the ethic of the first Puritan colonists, the birds would have been shot as pests. Subsequently, their beautifully colorful feathers became used for adornments, particularly of women's hats. The birds were so persecuted that by the time of the Civil War, the species, once widespread east of the Mississippi River, was scarcely found outside of Florida. As with several other extinct species we have discussed, flocking parakeets were particularly easy to kill, especially since flocks would frequently return to the spot where others of their kind had just been shot. The last wild Carolina parakeets survived into the early 1900s. Ironically, the last captive bird died in 1918 in the same cage that had housed the last passenger pigeon four years earlier.

The concept of a "pest" species is in the eye of the beholder. We discussed that earlier regarding the Kolam, a design made in front of homes in India with rice powder to attract ants and other insects. Widespread Indian values do not endorse the killing of animals, even those that are pests. That is apparent immediately to anyone who visits India. Other factors, too, impinged on the survival of the Carolina parakeet such as destruction of trees containing the cavities it requires for nesting. Nevertheless, I expect few ecologists would doubt that the Carolina parakeet would still adorn the landscape of the eastern United States had the bird been tolerated by Indian colonists, rather than persecuted by Puritans.

Four other bird species have become extinct in the United

States since the landing of the *Mayflower.* These are the Labrador duck, dusky seaside sparrow, Bachman's warbler, and the famous ivory-billed woodpecker. All of these succumbed to extinction either due to human disturbance, destruction of their habitat, or to causes not entirely clear.

Extinct birds of India

Remarkably, this will be a decidedly brief discussion because over India's long history, only two bird species are believed to have become extinct. One of these is the pink-headed duck. Chandley (Avibird website 2021) states that "Even before the species decline, the pink-headed duck was always a shy and secretive bird by nature. And research reflects that they may have always had a small population size. The cause for their decline is still somewhat a mystery. The main threat to the species is believed to have been the destruction of its habitat for agricultural land. The impact of invasive water hyacinth may have also been a contributing factor." Colonial sports hunters also impacted the bird as did some food marketing, but the latter factors were not perceived to be of major importance.

The second bird believed extinct in India is the Himalayan quail, a bird of the Western Himalayas which occurred at altitudes over a mile high. First discovered in 1846 and known from only two locations and 12 specimens, the bird has gone unrecorded since 1876, thus little about it is known. One thing that appears certain, however, is that it ranged far above elevations inhabited by Indian communities. This, and the bird's isolated and very limited distribution suggest that whatever the causes of its demise, the impact of Indian communities was not likely a major contributing factor. In sum, only one bird species of broad distribution in India is known to have become extinct and the cause of its demise was likely habitat destruction.

Extinct Birds of the United States and India

United States	India
Heath Hen Passenger Pigeon Great Auk Eskimo Curlew Carolina Parakeet Labrador Duck Dusky Seaside Sparrow Bachman's Warbler Ivory-billed Woodpecker	Pink-headed Duck Himalayan Quail
9 Species	**2 Species**

Observation

What we have seen by analyzing eastern North America's extinct bird species is striking to say the least. Of the eight extinct species for which we have adequate information, five were driven to extinction primarily due to over-hunting. In comparison, India has lost but one bird species of broad distribution for which the causes of its demise are unclear. Hunting was a practice the Puritans and their descendants found totally acceptable. Contrarily, had the *Mayflower's* passengers been emigrants from India, such would not have been the case. Few or none of these birds would have been killed over decades if not centuries for food, feathers, sport, or as pests. Consequently, far more likely than not, they would all be with us today, and some, particularly the passenger pigeon and great auk, might still be viewed from the vicinity of Plymouth Rock, just as the first emigrants would have seen them in 1620. Yale professor Steven Kellert (1986) reflects precisely on these observations. "The study of vanishing wildlife is necessarily the study of man's perceptions of animals. What we fear, what we hope, and what we admire in animals will inevitably determine their fate."

We have looked only at birds in this discussion. That is because they account for most of the extinct species in recent times and their histories of decline are much better known than of other taxa. Nevertheless, our earlier reviews of major predators, the top herbivore, as well as poisonous snakes resulted in exactly the same outcome – in India's favor.

But wait!

What About The Wild Turkey?

Perhaps nothing is more closely associated with the Pilgrims than the gobbler on which they reputedly feasted during the nation's first Thanksgiving. That once over-hunted gamebird did not go extinct. Does not the wild turkey's restoration serve as a counterweight to demonstrate the success of US conservation practices? Is that bird not an outstanding example of a conservation success story? Well, in some ways it is. It is one of the most outstanding examples of species recovery in the United States.

As with many of the birds discussed in this chapter, the wild turkey in the United States had declined precipitously until the beginning of the 20th century, primarily due to over-hunting and cutting of forests for timber. By the beginning of the 20th century the species had been eliminated from 18 states and approximately 30,000 birds survived. Protective legislation, specifically the Lacey Act and the Migratory Bird Treaty Act, were important first steps in conserving this and many other birds that were in decline. Around the same time, efforts were made to restore turkey populations by releasing pen-reared birds. Overall, these efforts failed due to the birds being too naïve to survive in the wild. Subsequently, in the 1940s, when wild-caught birds were transplanted to sites from which wild turkeys had been eliminated, these predator-smart birds thrived. Primarily thanks to the Wild Turkey Federation, over 200,000 turkeys were transplanted throughout regions where they had previously occurred and now the bird has been restored in every state, the population having increased to over 6,000,000 birds (National Wild Turkey Federation website 2021). Presently, in some localities, the birds are so common as to be a nuisance by some.

Does this success counterbalance our failure with those species we drove to extinction? Hardly. A large asterisk should be placed next to this success story. The asterisk is based upon the motivations behind this achievement. This effort was undertaken because the wild turkey is a gamebird – a bird hunters love to shoot. Most other wildlife conservation success stories in the US fall into this same category – the white-tailed deer, the elk or wapiti, numerous species of ducks. The hunting community deserves credit for these successes. However, saving animals primarily because you want to shoot them is hardly an ideal conservation value. For one thing, it sets up a powerful hierarchy among animals – game-animals, non-game, fur-bearers (animals to be trapped), vermin, predators – and being that the vast majority of our wildlife are non-game species, they receive the short end of the stick – and then some.

As a consequence of early conservation efforts in the US revolving around over-hunted species, both federal and state fish and game agencies, when first created, were built around this nexus. This relationship was strengthened in 1937 when the Federal Aid in Wildlife Restoration Act was passed, and later, in 1950, the Sport Fish Restoration Act. These acts created excise taxes on the purchase of hunting and fishing equipment such as guns, ammunition, fishing rods, and the like. Revenues from these acts, which amount to many hundreds of millions of dollars per year, are, by law, distributed to the states based on the area of the state and its number of hunters. As a consequence, many state fish and game departments were financed, especially in their earliest days, mostly, if not entirely, by these funds.

The parody of the Golden Rule, "Whoever has the gold makes the rules," fits perfectly in this context. State fish and game agencies have been controlled by hunting interests despite the fact that this pastime has declined in the United States to the point where only about 5 percent of the populace hunts. Recognizing the tenuousness nature of such small minority control, coupled with a general decline in hunting as a hobby, has fostered a relentless power struggle, and a decidedly effective one, for this old-guard – the staunch supporters of the North American Model of Wildlife Conservation we shall touch upon later – to retain control.

"Power doesn't corrupt people;
people corrupt power." – WILLIAM GADDIS

Some will say that the circumstances I describe are a thing of the past. While the situation is gradually improving, the abundance of wildlife killing contests allowed in nearly every state, suggests that the states still have a very long way to go. This has major implications and is one of many reasons why conservation needs to transform its approach.

CHAPTER 5

A MORE EFFECTIVE PATH TOWARDS ACHIEVING CONSERVATION

The insights derived from our discussion to this point suggest that the traditional presumptions of many wildlife conservation efforts in the US and elsewhere are seriously flawed. They fail to focus on the most important single factor influencing whether effective conservation will occur or not – the value systems of the people concerned. The consequence of this flaw is that the programs and initiatives developed to achieve wildlife conservation are much less effective than they could be.

The approach we explore here is built upon our understanding that the basis of effective and enduring conservation is the individual's – and ultimately a society's – values system. The more that conservation values are part of that makeup, the greater the potential for effective conservation action to take place.

Our values system represents the essence of who we are. It is our emotional core. Importantly, extensive research demonstrates ever more conclusively that it is our emotional being, what we feel in our hearts, which determines the decisions we make (Winter and Koger 2004; Clayton and Myers 2009; Heberlein 2012). These findings are at odds with the long-standing, dominant approach of the conservation community which focuses on the procurement and presentation of conclusive hard biological data as the basis for all decision-making. Lamentably, while many conservation professionals are out doing field research on wildlife, the species and habitats we seek to understand are spiraling on a downward trend.

To begin to understand this proposed approach, we can learn from the shortcomings of the present emphasis on biological data collection reflected in the ever-increasing study of what are referred to as neotropical migratory birds. These are birds which breed in the US and Canada, but migrate south of our borders to spend the winter. Some migrate only to Mexico and the Caribbean, others as far south as Chile and Argentina. Neotropical migrants include hundreds of species, many common and known to most of us, such as the robin, bluebird, and meadowlark, not to mention species popular with hunters including the ducks and geese. As interest in these birds ramped up in the US in recent decades, numerous initiatives were created to conserve them.

There was good reason for this. Most importantly, increasing numbers of species, approximately one of every four, had declining populations. For a few species, the cause was known, for many it was not. Because, by definition, North American neotropical migratory birds spend the winter south of our borders, it was very possible, if not likely, that in some cases their decline was due to conditions on their wintering grounds abroad.

Given this circumstance, what would seem the most practical approach for dealing with a problem of this sort? Does it not appear the essential first step would be to establish a strong, inclusive mechanism for communication and collaboration with institutions in the countries where the birds winter? This would enable us to learn what is known locally of their status, threats, where the species fit in the conservation priorities of the country, what is being done on their behalf, and so on. Such a dialogue would result in the development of respect and trust, both parties understanding one another's priorities, and potentially the development of a joint work plan of mutual benefit.

While this might seem the most practical approach, only one of the various initiatives created in recent decades proceeded in this manner. In nearly every instance, a group of US "experts" met, with virtually no representation from the countries where these birds winter, and applied their collective wisdom to develop a plan to conserve the birds in question, including on their wintering grounds. Follow-up abroad to these plans basically entailed figuring out how to get other countries (whose language in most cases we do not speak and whose customs we do not know) to achieve the objectives we had set. Such was the case for ducks and geese when the North American Waterfowl Management Plan was developed. The same was so for Partners in Flight which aimed to conserve migratory landbirds. Conservation of shorebirds took a similar tack via the Western Hemisphere Shorebird Reserve Network. The Neotropical Migratory Bird Conservation Act, legislation to conserve all species of neotropical migratory birds, followed a slightly different scenario, but one even more off-base. In that case, a deliberate effort was made to **exclude** participation of representatives from the countries where "our" birds winter. I offer this assessment based on close, first-hand experience with all of these initiatives.

Typically, such initiatives, as seen above, focused heavily on conducting research to document the status of each species, or in having other countries designate protected areas for the species of concern. (As we saw with the monarch butterfly, site designations often mean little and, in some cases, can be counterproductive.) Landbirds, shorebirds, waterbirds, seabirds – you name it – each initiative used variations on the same broken model.

It was believed that the results from such efforts conducted in the countries south of our borders would strengthen conserving these species, both in the US and abroad along with the habitats on which they depend. It was a shock, therefore, in the case of the landbird initiative when research determined that many of these neotropical migrants depend more on cut-over secondary forests for their wintering grounds than they do on mature forests. To the regret of the researchers, much of the data collected actually could be used to support the cutting of mature forest rather than its protection! Bad news! Researchers had to do a fair amount of back-peddling to wiggle out of this mess.

Had these initiatives centered instead on immediately engaging other countries, on understanding their perspectives, priorities, and societal values as they relate to wildlife and what is being done on its behalf, I have no doubt that advances to conserve neotropical migratory birds would be much further along than they are today.

What should serve as our ultimate conservation goal is not at issue here. At issue is the path we choose to get there – by what mechanism we strive to achieve such goals. Whether our ultimate goal is recovering endangered species, preserving disappearing habitat, sustaining migratory bird, bat, or fish populations, or whatever – the approach to achieving those goals is what matters. That is where a major overhaul is necessary in how we approach conservation.

The Slow Development of Values-oriented Conservation

Developing a set of conservation values is not a new idea. Well over half-century ago, Aldo Leopold wrote that conservation could not effectively be achieved without "an ethic dealing with man's relation to land and to the animals and plants which grow upon it" (Leopold 1949). But, Leopold gives no indication about how to create an ethic (Heberlein 2012) suggesting that because such an ethic socially evolves, "nothing so important is ever written … the evolution of a land ethic is an intellectual as well as emotional process" (Leopold 1949). Such an approach does not take advantage of the importance and power of enunciating a shared vision. The omission of a how-to road map for creating this ethic and validating it in some way, in the words of Heberlein (2012) "leaves us on our own, and there we struggle."

Woefully, Leopold passed away prematurely, short-circuiting

the potential to expand his thinking on the matter and refine just how such a goal might be achieved. It is time that we pick up where Leopold left off for, as Heberlein (2012) so correctly points out, "the process has lacked direction and a map" (2012). Stated more assertively by conservationist Matthew Child (2009), "So why isn't there a culture of conservation? Because there is no ideological precedent to establish the culture, not even within the conservation movement."

This is not to say that conservation values have been entirely ignored. The issue has undergone substantial development, but primarily in one isolated area – academia – where, regrettably, it is decidedly disconnected from most on-the-ground conservation efforts. Academicians take great pains to break down conservation values into various types, such as existence values, amenity values, option values, transformative values, among others (Perlman and Adelson 1997). Others separate them into spiritual, use, lifestyle, place, and ethical values (Infield and Mugisha 2010). Academic journals too, such as *Environmental Ethics, Human Dimensions of Wildlife and Ecopedagogy* treat this field in minute detail. Here, we are specifically concerned with transformative or ethical values – those relating to an individual's philosophy of life. Minute detail is not our interest here. This is a big picture discussion relevant to everyday conservation practice.

All too large a portion of the US professional conservation community, despite glorifying Leopold, has given lip-service to his message and, if anything, has moved in virtually the opposite direction. The present conservation emphasis on payments for ecological services is a telling example of this movement away from Leopold's vision. The latter is an effort to assign a financial value to every aspect of nature from a black-bellied plover to the recycling of nutrients in a forest. The intent here is to defend conservation using economic arguments that decision-makers favor and understand.

The problems with this are many. Among them are that placing a monetary value on most things in nature is not easy and requires substantial investments of time and research; such monetary values can be quite arbitrary and easily challenged; many things in nature may have little or no monetary value; for many species and places, their monetary value cannot match that generated by other uses; and, the monetary value of nature usually belongs to a community which in many cases does not recognize this value and so decision-makers ignore the matter. As to Leopold's comments on such thinking, "The last word in ignorance is the man who says of an animal or a plant: 'What good is it?'" (Leopold 1949).

Even more extreme, a new brand of environmentalism, a post-preservationist view of the world, argues that we should "actively embrace our responsibility as shapers and builders of the planetary future" (Minteer and Pyne 2015).

It is crucial that the conservation community recognize that our deteriorating environment is a socio-political problem resulting not from lack of knowledge, monetarizing nature, or, as the inevitable consequence of human progress, but instead from society's development being driven by entities lacking constructive values (Herrera et al. 1976). High among these is the corporate sector which has long been driven by its bottom line – profits. Though this may be changing, and there are ever-increasing numbers of companies becoming environmentally concerned, this remains the exception rather than the rule. It is time that the conservation community accept and utilize the powerful tools created by the corporate sector and apply them in a positive way to a goal so meaningful to society. Fortunately, because conservation values help create meaning and fulfillment in people's lives, their acceptance and implementation should prove much easier to achieve than is the case for getting a society to embrace many of the valueless products promoted in the marketplace.

Learning from the Corporate Sector

I have touched on the downside of the corporate sector. There is also an upside. Corporations are powerful. They have moxie. They oftentimes have CEOs and boards with constructive visions which extend beyond profits and recognize the importance of sustainability. These must become major players in making this initiative a reality. The non-governmental sector of the conservation community including The Nature Conservancy, Ducks Unlimited, World Wildlife Fund, the Wildlife Conservation Society, various land trusts and numerous others, have already been quite successful at engaging this sector. But this has been more for specific causes, not the promotion of core conservation values. A shift must be made towards the latter.

Major advances within the corporate sector are reflected in the program of Corporate Social Responsibility (CSR), a measure of the social and environmental footprint created by a company. Corporate Social Responsibility began as a voluntary, self-regulated tool serving to provide the public with, in the words of the Business Directory, "a company's sense of responsibility towards the community and environment (both ecological and social) in which it operates." Such reporting aims to enhance a company's brand and reputation. The term was coined back in 1953 by economist Howard Bowen (Bowen 1953), but the practice did not really catch on until the present century. Since then, it has increased rapidly and at some levels has become more mandatory. As a result of the coronavirus (COVID-19) pandemic in 2020, interest in CSR has increased exponentially (Sharma 2020). Relatedly, the Corporate Eco Forum (CEF), founded in 2008, is an organization of Fortune and Global 500 companies committed to sustainability as a business strategy. Its combined industry revenues surpass $4 trillion.

Failings of the US Conservation Community

The US conservation community has been relatively ineffective at getting citizens on a broad scale to support and demand powerful conservation goals. A 2017 survey found that environmental causes received only 3 percent of charitable donations in the US, the smallest amount among any of the 9 primary charitable categories (Charity Navigator website). That a problem exists and calls for being fixed is not enough. A great debate is raging within the US over climate change. Is it human caused? Is the problem real? This is not a serious debate among scientists. Knowledgeable scientists offer a very broad consensus on the matter – climate change is real and is significantly influenced by human activities. Despite this, some highly-educated and powerful individuals who are non-scientists, particularly in the political realm, consider this phenomenon to be a hoax, or "that the jury is still out." Beyond that, they have waged a relentless campaign, absent of any accurate science, to discredit such a conclusion. Their efforts have been so vigorous and successful that the potential to achieve any meaningful legislative action on the matter has for long been nil. A long-term Gallup poll surveying people's attitudes about the impact of global warming found that in 1998, 31 percent of those surveyed believed that the seriousness was generally exaggerated. By 2014, with the potential of the threat being much more effectively documented, the percentage of individuals who felt global warming was exaggerated rose substantially to 42 percent (Gallup Poll website). Reflective of this, the overwhelming threat of climate change was a non-factor in all recent presidential elections and is ignored by Congress. The new Biden administration has pledged to reverse this, but the challenge will be daunting. Is this limited acceptance of the threat climate change poses due to a lack of data, or information? No. It is for lack of the issue hitting us in the gut, of it touching our hearts and emotions. The fallacious argument that climate change is a hoax created by liberals and big government – an emotional appeal by those who oppose action – has trounced the growing piles of data that demonstrate the problem is real.

Increased doubt regarding climate change

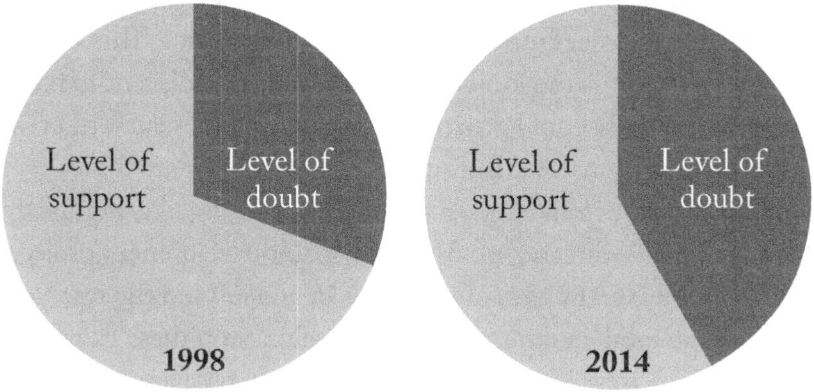

Changes in Attitudes Towards Climate Change *Source: Gallup Polls*

Emotions usually trump reason. That is a powerful aspect of how the human mind works and thus plays a central role not only in conservation, but any human endeavor. Still, in practice the concept is widely ignored (Lakoff 2004; Olson 2009). Cleverly remarked on by Andy Rooney, the famous television commentator, "People will generally accept facts as truth – only if the facts agree with what they already believe." Too many conservationists fall back on the safe-haven and false sense of professionalism found in the comforts of wildlife research and data compilation, though this all too often results in delaying to the future any meaningful action, because few, other than our peers, care much about our findings. There are books and reams of scientific studies which delve into this matter, but are outside my field and beyond our scope. In summary, we are still in what John Kotter and Dan Cohen (2002) describe as the analyze-think-change mode when we long ago should have shifted to the see-feel-change approach.

The "Nicaragua Canal Syndrome"

Internationally, there is a virtual mantra regarding the importance of working with communities in and around the world's protected areas and nature reserves to incorporate their cultural values into conservation efforts. Many hundreds of conservation projects focus on these communities. This makes sense. After all, it is the local communities in and adjacent to parks which are in daily contact with the living resources such reserves are set aside to conserve. But nature reserves today are under ever increasing threat from dams, roads, mines, and the like. Perilously, this trend is one-directional. Are these major development projects created and driven by the communities in and around the reserves? No. Can the local communities by themselves stop these externally driven enterprises? No. Developments of this type are decided in capitols by power brokers who more likely than not have never even seen the nature reserve which their project is about to impact. Such players, obviously, are of great importance in the conservation equation. Yet, you can probably count on one hand the number of conservation initiatives focused on these critically important individuals, a group inadequately engaged as a key conservation constituency.

This a reflection of the fact that locally-focused conservation projects do not and cannot exist in a vacuum. As successful as they might be, decisions made on a larger scale, such as by a national government, can decimate their successes overnight. I refer to this as the "Nicaragua Canal Syndrome" because I recall sitting, deeply surprised, in a board meeting of an international conservation organization while members expressed shock and dismay upon learning that Nicaragua was seriously considering building a canal to rival that in Panama, a canal which would, in one instant, eliminate some of the fine local conservation work they were doing. Somehow they thought things like this just could not happen – not to their projects anyway.

My experience in Puerto Rico previously had braced me for the Nicaragua Canal Syndrome – in spades! One of my first challenges as a biologist for the PR Department of Natural Resources was a proposal to create a model garbage dump, politely called a sanitary landfill, in a remote nature reserve. It just so happened that this reserve was the last refuge of the Puerto Rican whip-poor-will, a highly endangered ground-nesting bird that would have been subjected to marauding dogs, cats, and mongoose, likely causing its extinction. Where did this proposal come from? The Capitol, not the local residents. My next big challenge was a proposed copper mine in a reserve which supported the island's endangered broad-winged hawk. Then came the suggestion that a super port for oil tankers be built in the Mona Island reserve, home to the magnificent but threatened rhinoceros iguana. Then came the idea of a jail in another reserve, a highway through another, and communications towers in another. All of these threats were generated in the Capitol – none by the local communities residing around the reserves.

I use Puerto Rico only as an example of a phenomenon rampant across the globe. Tanzania recently considered construction of a road bisecting Serengeti Park. Australia is building an airport on an international protected wetland. Paraguay is developing Asuncion Bay, a locality protected under numerous conservation designations. In the US, there has been a relentless push, widely publicized, to exploit oil in the Arctic National Wildlife Refuge, one finally approved under the Trump administration. Numbers of US national wildlife refuges already have active oil and gas exploration. There are hundreds of other examples, no thousands. The Protected Area Downgrading, Downsizing, and Degazetting (PADDD) website documents over 3,000 such cases. These reflect decisions made in capitols beyond the influence of local communities. Decisions at higher levels are dominating interests at lower ones. In other words, national capitols overriding local communities. Ultimately, when the populace as a whole is disinterested in nature, their politicians are as well, resulting in consequences that are nothing but devastating. Add to this the immense influence of lobbyists, many of whom are antagonistic to conservation, or have their own narrow view of it, and the problem becomes that much more severe.

The professional conservation community, in general, is far too narrow in its perception of the stakeholder groups it needs to engage with to be more effective. There are salient exceptions to this, such as the work of Bluestem Communications which focuses on values and identifying new constituencies. It does this on a project-by-project basis with an emphasis on the Mississippi River, Great Lakes, and upper midwest. The organization WildAid does an incredible job on this front regarding illegal wildlife trafficking. But, lamentably, such efforts are scarce and underfunded. The conservation community generally would benefit from broader thinking in this regard.

Do Values Change Too Slowly to Make a Difference?

Most of my associates in the conservation field with whom I have discussed the importance of values seem to believe that changing or developing values requires a lengthy process, perhaps across generations. How often is it argued, "Oh, but values take so long to change. We don't have the luxury of time. Values might be a long-term goal, but we have other pressing priorities of more urgency." Such a perspective has little chance of moving conservation forward.

We saw that change relating to values could occur rapidly in our examination of the Saint Lucia parrot conservation effort. Malcolm Gladwell (2002) dedicates his insightful book, *Tipping*

Point to explaining how change does not occur gradually but rather in dramatic fashion very similar to how a viral epidemic occurs, while Heath and Heath (2010) in *Switch* offer fascinating insights into creating change as does McKenzie-Mohr's (2011) *Fostering Sustainable Behavior.* It seems evident that the ability to influence people's attitudes and values is becoming easier every day. This is due to recent vast advances in communications technology coupled with the masterful development of well-conceived marketing. Over time marketing has become an incredibly successful promoter of dreams. To not accept this reality and incorporate social marketing as one of the most important instruments in our conservation toolbox is sheer folly. To be clear, I am not referring to environmental education. Numerous studies have demonstrated the latter not to be effective in changing attitudes and behaviors (ibid.). This is because environmental education focuses heavily on providing scientific information, not on addressing underlying values.

A University of California at Los Angeles study found that in 1997 the top five values emphasized in television shows, presumably due to their popularity among children, were: community feeling, benevolence, image, tradition, and self-acceptance. By 2007 the top five values were: fame, achievement, popularity, image and financial success (Dvorak 2013). Only one value – image – remained among the top five a mere ten years later. Of equal significance, look at the shift in the types of values being promoted.

Top 5 values emphasized in TV shows

1997	2007	2017
Community Feeling	Fame	Self-acceptance
Benevolence	Achievement	Financial Success
Image	Popularity	Image
Tradition	Image	Popularity
Self-acceptance	Financial Success	Community Feeling

Values Emphasized by TV Shows *Source: Varghese and Uhls 2021*

Though many in the conservation community have been reluctant to embrace the importance and potency of influencing people's attitudes and values, the corporate sector has seen benefits to doing just that. America's businesses reportedly spent over $240 billion in advertising in 2019 (Guttmann 2021). This is not money down the drain. Such enormous levels of funding simply reflect the widely-accepted sentiment in the business world that smart marketing will more than pay for itself. If this were not the case, just imagine how many fewer advertisements we might see! Perhaps the epitome of this approach is the cost of Super Bowl advertising. Imagine paying $5.6 million for 30 seconds of television time! Yet, doubtless every one of the businesses which invests in this extravaganza expects to be repaid handsomely by football aficionados who now have discovered a product they just cannot be without. In fact, Super Bowl advertising has become so refined and powerful that increasing numbers of people apparently watch the game not so much to see the contest as to revel in the new commercials which are specifically produced for this event. The conservation community may not have this kind of money to spend on influencing attitudes, values, and behaviors, but it needs to embrace this approach rather than spurn it.

This advertising – on what is it based? Madison Avenue

abandoned fact-based marketing back in the 1950s, replacing it with an identity-based, storytelling approach. The consequence was no less than a complete reversal of a core American value from one of thrift to one of consumerism (Sachs and Finkelpearl 2010). Marketers, in ever more refined efforts to influence the public, have embraced the new psychological and neurological data on how our minds work, our unconscious, as well as our conscious, data which has increased exponentially in recent years. As a result, they have developed techniques such as metaphor elicitation, consensus mapping, response latency, priming, implicit association testing, and neuro-imaging (Zaltman 2003). They may better understand what makes us tick than we understand ourselves. A broad spectrum of the conservation community, particularly, but not exclusively, governmental agencies, inadequately uses this renaissance of knowledge about the human mind and continues to embrace data and information as its primary tool for reaching the public. This is a tool which fails to recognize that we humans do not think in a linear, rational way, thus resulting in obsolescence and failure.

A most telling statistic has to do with influencing the values of our children – the segment of society most malleable in this regard. In 1983, corporations spent on the order of $4 per year on each and every child in this country to influence their wants through advertising. By 2010, that amount had soared to nearly $600 per child (Linn 2010) and today it is over $1,500. Imagine – for a family with three children that amounts to nearly $4,500 being spent per year to convince their kids about what they should pressure their parents to buy. No wonder so many parents today feel besieged by their children! Though some countries restrict such advertising, while Norway and Sweden outlaw it, such is not the case in the US where there are no restraints. Dr. Juliet Schor (2005) elaborates on the negative consequences of this in her book, *Born to Buy: The Commercialized Child and the New Consumer Culture.*

Corporate Advertising Per Year Per Child (US)

1993	2010	2020
$4	$600	$1,500

Corporate Advertising in the US Focused on Children over Time

Who should decide our values?

Values are inherent in our culture, in every culture, and each culture has its own. In the US, the conservation values of society's majority have been imprecisely enunciated or codified. The same can be said of most communities within the county. As a result, such values have never evolved into societal norms. Instead, what have been enunciated, codified, and heavily promoted are the values of the country's hunters and anglers. This, despite that fact that they make up 5 percent and 10 percent of the populace respectively. I have touched upon this elsewhere and will not delve into this issue here. Suffice it to say, hunting and angling have their place in US society. At the same time, that they should dominate conservation practice in this country in what I would argue is a decidedly undemocratic way, is another issue, one in major need of an overhaul.

This is not a favorable circumstance. While academicians debate the rate at which society's ethical precepts and values change, the marketplace is bombarding us with carefully crafted advertisements aimed specifically at influencing consumer choice – making some product, heretofore unheard of, become a necessity of everyday life. Does wanting a single-family home with a green lawn make a values statement? Certainly it does. Add on a nice outdoor grill, a backyard pool, a larger vehicle. We do not need a collection of surveys to tell us that American society's values are changing faster than at any time in our history. The same can be said for most modern societies. The direction of the change – consumerism.

I was fascinated by a TV show in which three different households each undertook a $100,000 makeover of its backyard patio. Needless to say, there are many ways to improve a patio with $100,000. Imagine all the beautiful stone paths, the spectacular plantings, enhanced wildlife habitat, perhaps a pond thoughtfully created and landscaped to be neutral, if not a net positive, regarding its potential impact on resource use and climate change. Not so. The main determinant of which house had gotten the biggest financial bang for its buck? The size and design of the outdoor grill and associated paraphernalia that had been installed. The winner not only had the most impressive grill with a refrigerator readily at hand, but the coup de grace was the installation of not one but two 55-inch flat screen TV's next to the grill, placed at different angles so that the owner would not miss a single minute of football, no matter which way he turned.

What values are represented by this battle of the grills? The message should be quite clear. The United States' values are changing more rapidly every day – and not necessarily for the better. Does not excessive consumerism have an environmental price? Does not the dramatic ascendancy of values based upon consumption simultaneously precipitate a decline in values based upon the sustainable use of resources?

"Nowadays we know the price of everything and the value of nothing." - OSCAR WILDE

Just how much is being spent to counter this trend, to reinforce and build upon other American values which are more long-standing and could foster a sustainable planet? Where are the campaigns, backed by marketing strategies and branded messaging, which remind the public of our core conservation values and ultimately result in a better world for our grandchildren?

The natural sciences are a fabulous tool to help us understand the world. They can lead to all sorts of inventions which may either improve our lives or destroy them. They have no values associated with them. The natural sciences can be used for good. Or, they can be used for evil. Natural sciences can tell us what is. But they cannot tell us what ought to be. That is up to us to decide. Wisdom must come from another source. In the words of David Orr (2002), "The increasing velocity of knowledge is widely accepted as sure evidence of human mastery and progress. But many, if not most, of the ecological, economic, social, and psychological ailments that beset contemporary society can be attributed directly or indirectly to knowledge acquired and applied before we had time to think it through carefully." To paraphrase Nancy Miaoulis (2002) with the Institute of Global Ethics, in the end, it comes down to the values we commit to and the lens of ethics we apply to situations. Our over-emphasis on natural science studies must be modified. Our focus will have to be on something more intimate and lasting.

Sadly, the western world's major religions, while expansively explaining how people should relate to one another, as well as to their god, were typically remiss when it came to our relationship with the environment and all the living things which make the world such a splendid place. Many books and seminars have been devoted specifically to this topic. In a similar vein, our nation's founding fathers, as brilliant as their outline for democratic governance may have been, as well, totally missed this point of how we treat the world around us. It goes unmentioned in the US constitution and all its amendments.

As a result, we have some catching up to do. The era has come for US society, for all societies – from the smallest communities to the largest of governments – to seek a set of values, as comprehensive as possible, which address our relationship to the earth and the living things which cohabit it with us. This point was reflected upon by Aldo Leopold (1949) who noted the need to "extend our system of ethics from the man-man relation to the man-earth relation. We shall achieve conservation when and only when the destructive use of land becomes unethical."

Surveys have been conducted for years which help identify US conservation values. They provide a good starting point. "The country, however, lacks a clear, common vision for how much nature it wishes to conserve, in what form, at what cost, and for whom. As a result, the United States is vastly underutilizing its capacity to conserve nature" (Lee-Ashley et al. 2019). The need for coordinated framing and messaging to reinforce such values, may be the single greatest shortcoming of the conservation movement today. It is time this be remedied. After all, as stated by Lee-Ashley, et al. (ibid.) "These shared values, if channeled through available legal, advocacy, scientific, and political pathways, give current generations immense power to successfully curb wildlife extinctions, fight climate change, reduce toxic pollution, and safeguard healthy natural systems upon which future generations will depend."

The North American Model of Wildlife Conservation discussed earlier is reputed by some to contain concepts relevant to a discussion of America's conservation values. I, and many others, see the Model more as a descriptor, and a biased one at that, of the evolution over time of the conservation movement in this country. It is more a historical representation than a set of values statements. Worse, in the instances of many states, it has been hoisted as a false idol, a false set of values, representing the interests of the very few, at the expense of identifying truer values representative of the state's populace as a whole.

As a point of reference, a few countries, India for example, have incorporated a statement concerning conservation values into their constitutions. India's reads, "The State shall endeavor to protect and improve the environment and to safeguard the forests and wildlife of the country" (Article 48A). And, "It shall be the duty of every citizen of India to protect and improve the natural environment including forests, lakes, rivers, and wildlife, and to have compassion for living creatures" (Article 51A). In 2011, both Bolivia and Ecuador amended their constitutions to grant rights to nature.

India's Constitution

"The State shall endeavor to protect and improve the environment and to safeguard the forests and wildlife of the country."

"It shall be the duty of every citizen of India to protect and improve the natural environment including forests, lakes, rivers, and wildlife, and to have compassion for living creatures."

It is not essential that the US undertake such an amendment to our constitution, or that communities pass legislation relating to their values. What is important, as a start, as the first phase of a values-oriented strategy, is simply to seek out those conservation values which are most broadly accepted by a community so that they may be clearly articulated, shared, emphasized, strived for, and further developed. Identification of such values will enable recognition of a common purpose. Far too many issues highlight societal differences rather than compatibilities. Yet, it is in areas where people see eye to eye that the most progress can be made.

There is no need that conservation values be codified into law. It is not necessary that they be formally adopted in any way.

Informal recognition is fine. Their simple identification and general acknowledgment will go a long way towards helping guide future conservation efforts and give direction to the powers that be who guide each community's future. For one thing, this makes their continued development more flexible – a living document that can be improved regularly over time with increased attention to this issue.

Some will doubt that it is possible to identify broadly accepted conservation values in today's society given the polarization of the US political system. Perhaps they are right. All the more reason to begin at a smaller scale with local communities and civic organizations.

The four strategies recommended by The Social Capital Project (2008) to better improve reaching the public are all notably synchronous with and supportive of the approach we are exploring here. These include: (1) Redefining what it means to care about the environment; (2) Making environmental issues personal; (3) Turning personal actions into collective action; and (4) Filling people's need for social connectedness and a sense of purpose. Excellent – this is a fine piece of work. It provides useful advice for moving forward.

Social Capital Project Strategies

1. Redefine what it means to care about the environment;
2. Make environmental issues personal;
3. Turn personal actions into collective actions;
4. Fill the need for social connectedness and sense of purpose.

Clearly, any broad statement of values is open to interpretation. After all, what is meant by leaving the earth in "good shape" for example? I do not believe it is critical that such words be defined. As mentioned earlier, we are not writing a law. No one would be going to court over this. Some may choose to believe that throwing wastes into a river somehow does not put the earth in worse shape. That is their choice. What is important is that we would now have a widely-accepted common goal. We would have something quite concrete towards which to strive.

A comparable value in the political realm would be "freedom." Do not virtually all Americans believe in it? Yet, the *Richmond Inquirer,* representing what in the mid-19th century was a widely-held view, defined it as "not possible without slavery" (Oakes, 1998). Because freedom is referred to in our Constitution, our courts are still working to define this elusive term to this day. What would our country be like without such guiding principles as "freedom," "liberty," "democracy," and "justice"? It is time we had some noble equivalents which represent our country's conservation vision.

As indicated earlier, I do not believe conservation values need to be crafted from scratch so much as they need to be identified – and subsequently built upon. Some already exist. Our challenge is to find and to prioritize them so that those reflective of the broadest swaths of society rise to the top. How we might go about this is the focus of the next chapter.

CHAPTER 6
IDENTIFYING CONSERVATION VALUES AND REACHING CONSENSUS

The Individual

Before delving into community conservation values, let's turn, for a moment, to the individual – the fundamental building block of every community.

As great a challenge as any in achieving impactful conservation goals is for each of us to recognize our power as individuals to create change. It is only if we create high goals for ourselves that we should expect more from the communities of which we are a part. Ultimately, the extent to which any initiative will succeed hinges on the extent of buy-in by each individual community member. As noted earlier, conservation begins with each one of us. We need to be champions for what we believe in. If we do not take personal actions to address issues, we become part of the problem. That is because each of us depends upon the earth for practically every aspect of our lives, so how we treat the earth merits our attention and our actions.

Since each of us belongs to a number of communities (our local town, the civic organizations of which we are members, our professional societies), any one of us can take the lead to move forward the types of initiatives described in this book. Each of us can be the change we would like to see in the world, the biggest challenge is taking the first step. Below are some specific actions concerned individuals might take that can build individual empowerment and lead to the values-centered change we so desperately need:

"Everyone thinks of changing the world, but no one thinks of changing himself." - LEO TOLSTOY

"Be the change you wish to see in the world."
- MAHATMA GANDHI

1. Identify your personal conservation values and implement them through your behaviors: This is a personal self-examining, goal-setting exercise. It need not be a complex process, though that does not make it any less challenging since it is far easier said than done. Each of us has underlying sentiments concerning our treatment of the earth – presumably positive ones. We should reflect on these and, most importantly, convert our thoughts into actions. All too often, though we believe in something, we do not act on it. That is human nature. But we pay a price for this – for not making our voice heard. This element, exploration of our own personal conservation values, is important because conservation should be a bottom-up practice, one that starts with the individual and builds, democratically, into a pillar of a community. Further, it is individual behavior that influences group behavior. And group behavior creates societal norms – patterns of behavior that group members implement without question because the take them for granted.

"Never doubt that a small group of thoughtful, committed, citizens can change the world. Indeed, it is the only thing that ever has." - MARGARET MEAD

2. Promote the identification and implementation of conservation values in the communities (organizational, local, state, national) of which you are a part: This is self-explanatory. It is also simple and concrete. Nothing dramatic is required. Letting your concerns be known is a great start. The same for starting at a local level and with smaller, more homogeneous entities. How things evolve from there will vary considerably. The extent to which your sentiments are taken seriously might well influence your level of support to or engagement with that entity in the future. It is important to support groups that identify their values and take them seriously. One specific action might be to peruse the website of the community in which you live to see whether it expresses in any way its values regarding the environment. If it does not, it is time to start. If it does, what is being done to implement them? Can they be strengthened? Similar inquiry can be done at higher levels, particularly the state. Generally speaking, in the US, I believe state values concerning wildlife conservation lag far behind those at other levels, so there is plenty of room for improvement on this front.

3. Support groups with the specific mission of promoting conservation values: Such groups are very special. Keep them high on your list if they focus on the elements of change we have been discussing. Such groups address the roots of a problem, not the **symptoms.** This would result in most of them focusing on particular societal issues to address their conservation goals. Such groups might use innovate tools to reach their target audiences.

4. Favor entities that report on concrete steps taken to achieve their conservation values: There are increasing numbers of groups that strive to measure their achievements. Measuring success is of obvious importance. The challenge is to select the correct factors for measurement and to perform it accurately. The setting aside of protected areas, for example, used to be thought an excellent measure. It is easy to measure the number of new acres "protected." We discussed earlier why such a measurement is increasingly untrustworthy. Support groups that measure and report on how they have improved individual or societal actions.

5. Withdraw from or avoid supporting entities that reflect contrary values: This is simple with regard to organizations. It is more complicated in the case of businesses whose products we purchase. It is difficult to know, for example, the environmental values of the companies that produce what we purchase. Also, most of the products we buy do not convey information on the environmental impact of producing them. As a result, we let price be the deciding factor – not always a good choice if you care about the environment.

6. Become increasingly informed and involved: We should all strive to make the world a better place and that begins both with caring and being informed. Though society is increasingly complex, access to information is more accessible than any time in human history. It is the challenge of every citizen to understand the conservation issues affecting themselves, their community, and the environment generally, and follow that up with becoming engaged in some way to positively influence decision-making and potential outcomes. Ultimately, the more involved we are as individual citizens, the greater the potential for a better world around us.

"The most difficult thing is the decision to act, the rest is merely tenacity." – AMELIA EARHART

Individual Actions:

1. Identify your personal conservation values and implement them through your behaviors
2. Promote the identification and implementation of conservation values in the communities of which you are a part
3. Support groups with the specific mission of promoting conservation values
4. Favor entities that report on concrete steps taken to achieve their conservation values
5. Withdraw from or avoid supporting entities that reflect contrary values
6. Become increasingly informed and involved

It has been widely demonstrated that individuals possessing a set of values in no way guarantees they will generate behavior change in line with those values. For example, a person may oppose drunken driving (the value), but make an exception for him/herself and do it anyway (the behavior). Thus, translating our values into constructive conservation actions and behaviors adds an additional level of complexity to the challenge. The identification of values is an essential step – I believe the first. Meaningful individual behavior change cannot occur without it. Identifying and promoting specific **vital actions and behaviors,** like those described above, will serve in a major way to implement these values in everyday practice. Such is the case at both the individual and community levels.

"We've got to pause and ask ourselves: How much clean air do we need?" – LEE IACOCCA

Individual behavior change must be expanded upon by molding conservation values into social norms, starting with a focus on a few specific vital behaviors and building from there. Because humans are social animals, much of our behavior is a reflection of what others around us do, particularly individuals whom we hold in high esteem. For that reason, the establishment of societal norms is of great importance. As increasing numbers of individuals shift to more conservation-sensitive behaviors, those around them take notice and the shift accelerates. That is the power of norms.

Elephant Ivory – The Life Blood of a Small Connecticut Town

There is a meander in the Connecticut River where, over the course of millennia, the ceaseless current has etched away the underlying granite so as to create a deep basin abutting the shoreline. This natural feature enabled boats of deep draft to dock directly at wharves onshore to load and unload cargo – an invaluable asset for commerce. Adjacent to this landing, the region's contours further endowed the landscape with a creek which, when dammed and manipulated craftily, created fine conditions for mills and industries of one sort or another to power their machinery.

This commercial hub, which became the town of Deep River, was not unlike many such hamlets which sprung up in New England in the early days of our nation – but for one important exception. Deep River's primary industry for well over a century was the manufacture of ivory products. In 1798, Phineas Pratt invented a water-driven circular saw which replaced the cutting by hand of cow horn and bone for making combs and other objects. This invention, accompanied by increased trade with Africa, resulted in the rapid substitution of African elephant ivory for horn and bone to make these products. Ivory produced not only a better and prettier comb, but also a more durable one. In 1809, the first ivory carving factory opened in Deep River. It was not until 1954 that the last shipment of ivory arrived from Africa.

Over time, the products made from ivory shifted from combs, buttons, and billiard balls to piano key veneers. By 1910, it is estimated that Deep River and the adjacent town of Ivoryton were producing 90 percent of the piano keyboards manufactured in the United States. During its heyday, Deep River's largest manufacturer, Pratt, Reed & Company, was utilizing 12,000 pounds of ivory per month. Ninety percent of the ivory imported into the United States was used by this town. Its prosperity led to its moniker – Queen of the Valley.

ELEPHANT IVORY PRODUCTS FROM TOWN OF DEEP RIVER

The ivory used by the manufacturers in Deep River came from one place – Africa. And it came from only one animal – the African elephant. Estimates suggest that from 1850 to 1914 on the order of 44,000 African elephants per year were killed to supply this trade globally (Walker 2009).

But during much of the 19th century, the ivory trade was not just about elephants. It was also about slavery.

Zanzibar was the primary African port from which ivory was shipped to America. Zanzibar, though, is an island, actually a small cluster of islands to be precise. It has no elephants. Zanzibar was basically a transshipment point, its greatest commerce being in slaves. Merchants and sultans in Zanzibar would send massive expeditions to the African mainland to procure slaves for shipment to the New World. But why just bring back slaves? Why not have the slaves carry something of value on their trek to the East coast of Africa. That valuable material was ivory.

The journey undertaken by the enslaved African was no small matter. David Livingstone, the famous missionary and physician, estimated that five Africans died for every tusk brought to Zanzibar. That number is independent of the huge loss of life suffered by the slaves during their journey into bondage in the Americas.

Early in the development of the ivory industry in Deep River, it was deliberately misrepresented that the ivory came from dead elephants and that there was no slaughtering of these behemoths. Of course, that the trade had an association with slavery was never conceded either.

Ironically, while their businesses depended intimately on Africa's slave trade, both Julius Pratt and George Read were active abolitionists at home. In 1835, Mr. Read founded the Deep River branch of the Anti-Slavery Society and used his house as a stop on the Underground Railroad which aided in bringing escaped slaves to freedom (Walker 2009).

The irony of these conflicting values is extreme. On the one hand, we have Read and Pratt looking the other way with regard to the slaughter of massive numbers of elephants and the abhorrent toll on human life and dignity associated with the slave trade. On the other, we have their commitment to the abolition of slavery in the United States. How they justified in their own minds these conflicting positions will likely remain a mystery, but that is not the subject of our discussion. It is the town of Deep River on which we have our sights.

On April 24, 2013, approximately 100 years subsequent to the peak of ivory crafting in Deep River, the Deep River Historical Society voted unanimously to endorse "The concept of linking the history of the ivory trade in Deep River with today's wildlife cause of saving the African elephants through themed activities and programs, including cooperation with other community groups such as the Rotary Club and the regional schools." Two months later, on June 11th, the Deep River Board of Selectmen unanimously supported the resolution of the Historical Society and indicated its readiness to join with and assist the Society in its programs.

ELEPHANT STATUE IN TOWN OF DEEP RIVER

That fall, on November 9th and 10th, Deep River hosted a series of events including guest lectures on elephant conservation, viewing of the film *Battle for the Elephants,* student essay and art competitions focused on the elephant, and the dedication of an elephant statue in front of the town hall. Revenues from the events were donated to a respected African elephant conservation organization.

Deep River, a town which heavily owed its growth and prosperity for a century and a half to the crafting of ivory, had virtually reversed its position and embraced elephant conservation. That is to say, over time, its townspeople came to respect elephants as sentient, living things rather than perceive them heartlessly as a source of employment and profits.

There are likely a number of factors that facilitated this turn-around. African elephants were declared threatened in 1978 and importation of their tusks outlawed. Plastics had replaced ivory products in the marketplace. The public had become more aware of the elephant's plight. Regardless, this was a mega-change in the town's values – from pride in a unique industry built at the expense of elephants – to redressing a mistake others had made in the past to build that pride.

Importantly though, there was no pressure on Deep River to do this. No one was breathing down the town's neck to make recompense. It was done out of sheer good will. It was done to show that not only had the town's values changed, but that it wanted to back up that change by committing itself to aid in the salvation of the African elephant. By what seemed like a process of osmosis, the town's basic values had reversed.

And finally, it took an advocate, in this case Peter Howard, a local citizen deeply concerned about the African elephant's plight. Once Peter raised the issue, reaching of a consensus happened quite smoothly and efficiently.

The case of Deep River represents a very dramatic shift in values by an entire town. There was no urgency or profit to be made, no one was pushing. The town's action was purely a gesture of good-will based on caring for an animal which had been abused by earlier generations. What better example of a shared conservation value being tapped?

This is exactly what we need. We need a lot of Deep Rivers. But we cannot wait for such important initiatives to happen by chance. We need to accelerate the process. We need to expand it more broadly. The cultivation of shared conservation values that create a positive community identity needs to be promoted at every level from the smallest town to the entire nation – and beyond.

Deep River may well be the first town in our nation's history ever to pass a resolution to help conserve an animal on another continent. I would hope that it sets a trend for others. The process should be much easier when focused on domestic resource issues. Imagine tens, hundreds, thousands of Deep Rivers building on their shared values to conserve more effectively the US's natural heritage.

A Framework for Successful Community-level Conservation:

So, how might community values best be used to achieve conservation? I propose a framework with nine elements.

Framework for Effective Community-level Conservation:

1. Identify community conservation values
2. Create a charter of community conservation values
3. Frame, implement, and monitor conservation values
4. Focus on stakeholder groups
5. Customize cornerstone initiatives for each stakeholder group
6. Democratize all conservation practices
7. Network by expanding collaboration among communities at all levels from local to national
8. Develop and hire conservation professionals with socially-oriented skill sets
9. Adaptively manage and consider additional conservation values

FRAMEWORK ELEMENT #1:
Identify Community Conservation Values

The Community – Generally

The term "community" is used here in a broad sense to include not only a physical location such as a town, village, or municipality, but also communities of association such as clubs, civic groups, conservation organizations, professional societies, businesses, and corporations. After the individual, it is the community at its smallest levels, such as a town or civic group, especially those devoted to conservation, that provides the greatest potential for quick action and rapid advancement of the concepts proposed here. The relatively smaller size, greater like-mindedness, and simpler institutional structure of small communities will facilitate the development of a stronger suite of values in a shorter period of time. Identifying values in smaller communities also serves as a building block and provides valuable experience for implementing values identification in larger, more complicated communities. Perhaps most important, a groundswell of support from smaller communities that have already recognized their core conservation values will provide powerful leverage towards making state and federal efforts in this regard more rigorous, open, and successful. Importantly, there is little doubt that scattered communities here and there have already identified their conservation values. That is wonderful. But what we need now is a movement of each community, one by one, embracing its values, sharing them with others, and enhancing them over time to be that much better.

Identifying community values involves reaching a consensus on those core values most broadly embraced by the members of the community in question. Clearly, the task becomes more challenging as the size and diversity of the community increases. Though, consensus-reaching on a set of conservation values is entirely possible as long as there is goodwill within the community. The challenge is to identify those values that are most broadly embraced. Should it be found that even the most broadly embraced conservation values are less widely accepted than one might like, or that they are weakly held, the task does not become unachievable, it simply becomes more challenging – and more urgent. Such a result makes even more important a continuous effort to expand the community's thinking on this front so that, over time, its core values are strengthened and expanded.

Rancher Jim Stone of the highly effective Blackfoot Challenge, a collaborative effort led by western Montana landowners focused on stewardship of the 1.2 million-acre Blackfoot watershed bordering the Continental Divide, argues that the success of their program is its focus on the 80 percent of local people's interests that they all have in common, not the 20 percent on which they differ. For example, concerning human conflict with wildlife, cooperation has moved forward effectively on creating bear-resistant dumpsters and the fencing of beehives and calving grounds to exclude predators. Issues that might prove contentious, such as the killing of nuisance wolves or the poisoning of prairie dogs, would be left to other forums. Another rancher, Bill McDonald, insightfully refers to this area of common interest as the "radical center."

Why identify a community's conservation values? For a number of reasons. It is important to know the societal context within which conservation actions are being contemplated, because the more closely a proposed action can be related to an existing value, the easier it will be to achieve. Moreover, the process of identifying and becoming conscious of these values adds to their relevance. Importantly, once identified, the community can create measurable indicators reflective of those values and use these in assessing future actions and potential developments. And finally, identifying a community's conservation values provides a foundation upon which to build, to augment its conservation values over time. As Leopold (1949) insightfully recognized, "There is, yet no ethic dealing with man's relation to the land and to the animals and plants which grow upon it." and "I think we have here the root of the problem. What conservation education must build is an ethical underpinning.... Conservation may then follow."

Readers will have noticed, in the first portion of the book, that societal conservation values, in the US at least, have hardly been a model for wise conservation practice in the past. So, why should they now? Won't they, in many communities, reflect limited environmental concern, especially in the extraordinarily polarized political climate of today? The answer is a resounding "Yes!" But, think about the alternatives. The most obvious one is the one widely in practice, the "keep your head in the sand" option: Let's just ignore the issue of values and keep plugging away as we have been doing for decades and, if we keep working harder, and our hearts are in the right place, and we have hope, things will work out in the long run. The comment of Albert Einstein on this approach is apropos. Einstein's definition of insanity is doing the same thing over and over again and expecting different results. Ignoring a problem doesn't make it go away; things often get worse.

EINSTEIN'S DEFINITION OF INSANITY:

"Doing the same thing over and over again and expecting different results."

Another option is for an academic values-expert, an environmental author, or the conservation community at large, to identify a solid set of values and offer them as a comprehensive suite of goals for the general populace to embrace. Well, that would certainly be an easy solution – if it worked. But, such top-down approaches inevitably fail due to their lack of inclusiveness and insensitivity to local perspectives.

We should keep in mind a number of points:

First, conservation efforts only succeed when they address the root of the problem. Historically, great effort has been expended on knowing the status of wildlife. What are its numbers? Is its habitat stable? And so on. But, does this information focus on the true cause of the conservation issue at hand? Is it not more important to know the basic values of the human communities living in association with that wildlife? Is not the problem human-caused and values-based? Consequently, even if a community has only one shared conservation value, perhaps creating green-space in the community for future generations, this provides something to build on. If such is the case, it will likely be a very basic value and many issues facing the community can be connected to it. Success with that single value might well lead to subsequent identification of a second and a third.

Second, it is to fill such gaps that this effort should be ongoing. Though we begin by identifying a community's present values, we should also consider building support for future ones. Existing values should serve as a foundation. If we can demonstrate the utility of present conservation values in resolving issues facing the community, then building upon them will come naturally. Hopefully, over time, a community's shared conservation values will be strengthened and expanded. As environmental philosopher Bryan Norton (2005) points out, "Once this process starts, and there is an opportunity for communities to act… their action becomes gradually self-correcting."

Third, some gaps can be filled by identification of values at different scales – state, municipal, local, sectoral. This in itself has major positive potential as there invariably will be states, towns, municipalities, and communities which embrace more powerful and comprehensive conservation values than might be recognized in one's local community.

Fourth, the application of democratic processes alone will advance conservation even if the conservation values a community identifies cover little ground. Put another way, the process of fully and honestly engaging a community in a discussion of its conservation values has positive benefits, even if no conservation values of consequence are identified. As environmental philosopher Bryan Norton suggests, when we embrace democracy, and recognize its weaknesses, we can set out to educate and to improve the process, "seeking acceptable outcomes for now and continue amassing experiences as part of a truth-seeking process" (Norton 2005).

Fifth, what is being proposed here is not the abandonment of existing conservation efforts. What is being called for is a dramatic reorientation and reprioritization.

And sixth, if we look at what was a much more complicated and polarized situation where a focus on values worked – the case of South Africa's reconciliation process following the dissolution of Apartheid under the leadership of Desmond Tutu who won the Nobel Prize for his leadership – only two values were focused on during that process: "Truth" and "Reconciliation." No roadmap was drawn up. Nothing like it had been done before. The reconciliation of two seemingly irreconcilable adversaries was driven basically by these two values and a noble cause. At the time reconciliation was begun, only 22 percent of South Africa's populace was oriented by values to unite the country. Essentially, the outcome connected people who did not know one another to something greater than themselves.

"If you want to make peace with your enemy,
you have to work with your enemy.
Then he becomes your partner." – NELSON MANDELA

The bottom line: knowing where communities stand with regard to their conservation values, is a critical starting point. It provides conservationists, whether individuals or institutions, a base on which to build. It exposes the underlying source of all the conservation problems we face today. Sure, it is easier in many ways to address the symptoms – dirty air, contaminated water, increasingly endangered wildlife – but that will never solve the problem as long as underlying societal values are ignored.

Community values identified at simpler scales (civic group, small town) can serve as building blocks to identify values at more complex levels (state, nation). They can generate behavior change and action much more rapidly. Likely, they would encompass a broader array of values. This is especially the case for civic groups because of their greater like-mindedness. These are all positives. At the same time, identifying conservation values at more complex scales, such as state and national levels, are also important goals. States and nations need to strive to identify and implement values which best reflect all of their populace. It is highly possible that smaller scale initiatives will influence the future of conservation more than anything accomplished nationally, but this does not diminish the importance of more broad-scale efforts. Also, likely the most powerful approach to developing state and federal conservation values would be for the process to begin on lower levels – among conservation groups, in local communities, organizations, and civic bodies – then building momentum as it expands to the state and national level. Most practically, initiatives would begin wherever there are advocates who recognize that this must be done.

I do not mean to imply that no communities have undertaken identification of their values. In 2010, Pittsburgh became the first US city to recognize the legal rights of nature. Over and above that, it elevated the rights of people, the community, and nature over corporate rights (Lappe 2011). If a large, diverse city, such as Pittsburgh can take such a major step in this direction, it bodes well for communities everywhere. What is critical is that such efforts be stimulated in every community in the country, ultimately building into a national dialogue.

The Community – Conservation Donors

It is this community that has the greatest potential to transform the conservation landscape overnight. These entities, especially consortia of donors, such as the Environmental Grantmakers Association and Biodiversity Funders Group, have the financial where-with-all, the desire, and the flexibility to change the course of conservation for the better if they could only find the right path. I firmly believe these concerned brokers are constantly looking for one, but it has not been easy to come by. Hopefully, they will recognize the power of the framework proposed here. Earth Day Network, founded on environmental values, can play a huge role and, fortuitously, is seeking to promote values-oriented conservation globally. It is critical that others in the donor community do so as well.

Exactly what am I talking about here? I was invited to a meeting some years back by a major US foundation interested in capacity building, a theme I was involved in across Latin America at the time. Capacity building refers to enhancing the ability of local people, through various types of training initiatives, to manage their own affairs, in this case local conservation efforts. The discussion centered around what a new $100+-million global capacity building initiative might look like. We need the same discussion today, but focused on conservation values.

The Community – Non-governmental Conservation Organizations

If the suggestions in this book are meant for anyone, it is for conservation groups, large or small, whether local community land trusts or major non-governmental conservation organizations. This community already cares about wildlife and other natural resources, it is just looking for the best path forward. Becoming major players in helping other communities to identify their conservation values, set indicators for them, monitor them, reach out to youth, **and especially to help build stronger conservation values in the future** – these could be highly productive activities for any conservation organization. They must also collaborate better with one another, especially to develop and promote shared conservation values. I believe all conservation groups need to regularly reexamine what their mission is and, particularly, how best to get there – whether they are working towards it in the most effective way possible. Are they focusing on symptoms, the case in too many instances, or are they addressing underlying causes of the problems they hope to solve? Based upon the thousands of wildlife conservation proposals that passed across my desk, or with which I worked directly, it is evident there is plenty of room for improvement.

I'll offer you an example from the federal government because it is based on first-hand experience, but it pertains to the non-governmental community as well. The African Elephant Conservation Fund is run by the division I formerly managed. About $2 million were spent annually to promote the survival of this flagship species in the wild. This was clearly a worthy goal given the incredible intelligence and magnificent stature of these behemoths. As is widely known, poaching of elephants for their tusks has been increasing for decades to the point where ivory has become so valuable that organized crime using helicopters has become involved in the slaughter. Sickening images of dead elephants, animals mercilessly hacked to death with machetes or, nowadays, shot with machine-guns just for their tusks, are increasingly widespread in the news. So, what should be done about it? Obviously, everyone wants to deter the poachers. This can be achieved by hiring more guards, providing equipment, and training them better, among many other anti-poaching actions. Many of you receive solicitations for such needs all the time. What better way to save these animals than by keeping poachers from killing them?

Well, there is a better way. You find it by answering the question, "Why are the elephants being killed in the first place?" The answer is that elephant tusks are primarily carved into figurines that serve as a status symbol among the middle class in China. As this societal sector grows, and it is doing so rapidly, the demand for ivory is accelerating along with it.

So now, rethinking the issue, should all conservation funds go towards anti-poaching which inevitably gets worse as the price of ivory skyrockets? Or, should some significant portion of funding go to engaging the Chinese middle-class, a fair number of whom, 70 percent, according to a poll at the time, believed that the ivory being sold came from already deceased elephants? Perhaps a well-conceived mass media campaign alone could change their minds and reduce demand – thereby reducing poaching. Those were exactly the results with shark harvesting following a mass media campaign by WildAid to reduce the consumption of shark fin soup.

Well, there was no way I could convince elephant experts on my staff to dedicate funds towards demand reduction. In fact, when a draft of the executive order later signed by President Obama crossed my desk, a draft prepared by the State Department and reviewed by our division's experts, only one sentence in the draft referred to addressing the ivory market. The text was entirely about anti-poaching.

Though this example comes from the federal government, the big non-governmental organizations working on this issue do the same. They focus on anti-poaching. One reason is because it is easier for them to get donations from you for that purpose. No wonder the conservation movement is so far behind the eight ball!

Ultimately, until this sector transforms itself, conservation efforts will continue to struggle and lag further behind in an all-too-rapidly changing world. It is for this reason that I dedicate a subsequent chapter to training future professionals in this field. But, we can't wait for that. The professional conservation community must transform its way of thinking, and the sooner the better.

The Community – Youth

Of particular importance is the input of youth. After all, it is upon their shoulders and those of their children that the repercussions of our conservation actions or inactions, as the case may be, will primarily be felt. It is also they who have the greatest optimism and thus a willingness to explore outcomes many older stakeholders might not contemplate. To that end, it is especially important to engage youth organizations of all sorts in conservation values discussions. Youth will also be less defensive about many issues because they have not yet invested their careers promoting one line of thinking or another.

One could make the case that this entire values-setting exercise should be limited to the younger generation so that they might set goals for the communities, nation, and world they will inherit to manage. That is certainly an option as there are decided benefits to such an approach. At a minimum, youth should play a major role and, in many communities, could be the first to set such efforts in motion. Youth should also be among the first communities for which cornerstone initiatives are developed.

The Communities – Women, Minorities, the Underprivileged

Conservation of the environment and wildlife specifically across the U.S. and over most of the world has historically been almost totally dominated by men. And it shows. It happens, however, that a nation's wildlife belongs to all its citizens and women make up approximately half of them, minorities a substantial amount, and the underprivileged usually more than half. Further, particularly regarding broader environmental issues, it is members of these communities which bear an inordinate share of the negative consequences resulting from ineffective or inadequate conservation practices. The very limited engagement of these communities in wildlife conservation makes the field a very biased one if not downright elitist. While there are clear signs of movement in a positive direction to engage these communities, we have a very long way to go, and progress remains too slow. Implementation of the practices outlined in future chapters would dramatically improve advancement on this front. Youth should also be among the first communities for which cornerstone initiatives are developed.

The Community – State and Federal Governments and their Agencies

The identification of societal values has its importance at the state and federal levels for a number of reasons, some obvious ones being the power of these entities, the scale at which they deal with issues, and the precedent they set for the communities of which they are composed. Notably, it is the state agencies, in general, that are the furthest behind the times and in greatest need of overhaul.

The Community – Civic Organizations and Businesses

The homogeneity within each of these entities, coupled, generally, with their leaner organizational structures, will make the process of values identification that much easier. In most cases, it will also foster a broader suite of values than in larger, more diverse communities. In all likelihood, a number of these entities have already taken steps in the direction proposed here by developing a vision, mission statement, goals, and the like. Regardless, much more needs to be done.

Surveying Conservation Values

There is no single mechanism for identifying a community's conservation values. There are many potential ways to do this. And clearly, the larger the community, the more complicated the process. In the largest of these, such as state and national levels, the identification of those conservation values most widely shared among its populace should begin with conducting a review to learn from previous research and surveys which sought to identify such values in the past.

At a national level, the Biodiversity Project, now Bluestem Communications, has undertaken periodic surveys of the American public regarding how it values nature. The survey was conducted twice – in 1996 and 2002. The results are telling.

Participants ranked six values. Their level of importance was as follows:

Valuing Nature

	1996	2002
Personal responsibility to leave the earth in good shape for future generations	74%	56%
Nature is God's creation and humans should respect God's work	67%	55%
An appreciation of the beauty of nature	65%	54%
A desire to protect the balance of nature for you and your family to enjoy a healthy life	59%	48%
A desire, as an American, to protect our country's natural treasures and natural history	58%	48%
A respect for nature for its own sake	55%	47%

How the US Public Values Nature *Source: Bluestem Communications*

Much can be drawn from these data. Most significant is the steep decline in the importance of environmental values that occurred over the course of only six years. Accepting personal responsibility for leaving the earth in good shape for future generations, the most widely accepted value in both surveys, declined a dramatic 18 percent in that short span of time. On average, there was a 10 percent decline across the board regarding concern for the environment.

Replication of this survey was not conducted beyond 2002, but a separate Harris poll, performed in 2009 and 2012, addressed some of the same questions. The Harris poll included a question about leaving the earth in good shape for future generations. In 2009, the percentage of Americans concerned about this legacy was 43 percent. In 2012, it was down substantially to 34 percent (Steinberg 2012). In summary, between 1996 and 2012, a short period of 16 years, the percentage of Americans concerned about the state of the planet left behind for future generations declined from 74 percent to 34 percent. These results may seem unimaginable, except that both surveys showed significant declines in America's environmental attitudes on nearly every variable measured.

These surveys suggest the United States' environmental values have suffered a phenomenal decline in a matter of a few short years. If we choose to look, we can find the negative effects of this shift spread widely about us. For the present discussion, it is important to note three things. One, is that continuing to conduct business as usual – doing conservation the way we have done it decade after decade – has failed and, unless we change our approach, we can expect that things will only get worse. Second, the job before us will be that much more challenging if we don't address conservation values now due to this decline in our concern for the future. Third, this regrettable situation does not reduce the need to identify our nation's conservation values. To the contrary, it makes the need that much more imperative.

There have been other surveys. The Ecological Roadmap undertaken by The Social Capital Project is one. It finds, in line with previous surveys, a decline in the positive nature of our populace towards the environment. Others suggest that the public's concern for nature is not weakening (Doak et al. 2015). Should this be more reflective of existing circumstances, so much the better.

At any community level, a survey might serve as a useful fulcrum around which to launch a values initiative. Surveys provide us with, most importantly, a sense of how a community currently perceives conservation. How conservation fits into their lives, their thinking, their dreams for their future, and the world they leave for their grandchildren. From this information, it should then be possible to extract some basic conceptual values. These would reflect conservation principles found to be cross-cutting and widespread. They would likely be very general in structure. That is fine. In fact, it is perhaps better at this juncture that such be the case. It is via the next step that refinement and fine-tuning of the concepts would be achieved. Just how this is undertaken is central to the potential for broad acceptance and implementation.

Existing Values-focused Initiatives

The creation of a broad set of conservation values or ethics has not gone unattempted in the past. There are numerous examples at the local level, far too many to mention. One excellent case is the *Charter of the Chesapeake Conservation Partnership: Toward the Quality of Life of Communities Across the Chesapeake Bay Watershed* of the Chesapeake Conservation Partnership which includes at its core, eight shared principles to guide their partners collaborative conservation efforts in this large estuary on the mid-Atlantic coast of the US.

A noble effort was undertaken on a global scale which culminated in the creation of the *Earth Charter*, endorsed internationally by a wide range of organizations in the year 2000. The drafting process included representation from many countries of the world, including the United States. The document, comprehensive in scope, reflects a number of values not broadly held by American society today. It serves best as a source of potential elements and ideas from which to build a conservation ethic. Among other things, it calls for a new ethic of conservation and environmental stewardship and the need to mainstream them into conservation endeavors (Infeld and Mugisha 2000).

Why is the *Earth Charter*, for all the lofty intentions behind it, so little known and, though adopted by some cities and towns, remains unendorsed by virtually all conservation entities – local, state, and national – in the United States. Three major reasons are apparent. First, is the association of the Charter with the United Nations, which launched the initiative. Though the Charter later became a civil society effort, its early association with the United Nations gave it a stigma that some Americans hold in low esteem. Second, is the broad scope of the Charter. While it addresses fundamental concepts, such as caring for nature and living sustainably, it goes a big step further to promote the equitable distribution of wealth within and among nations. It speaks of demilitarizing national security systems and eliminating nuclear weapons. Such may be lofty goals, but they are ones that meet strong opposition among sectors of the American public. Third, the US conservation community, in general, still does not "get it" with regard to how central a set of conservation values is to achieving its goals. Too often, the importance of values seems to be an afterthought, rather than a central organizing principle within the conservation community.

It is evident, therefore, that for a set of conservation values to be broadly embraced in many US communities, or the nation as a whole, it must be a home-grown product and more restricted in its scope than is the *Earth Charter.* Being part of a global exercise does not seem to fit the US's psyche in this day and age. What is important is that each community identify its own comfort zone regarding conservation values. The community then has a shared goal towards which to strive, built on its own sense of place. It has a clear identity – a common vision.

In the 1990s, the US Forest Service created a land-use ethic. It is to "Promote the sustainability of ecosystems by ensuring their health, diversity, and productivity" (Thomas 1995). Though not lengthy, it covers a lot of ground. Also, had it been followed in the past, such an ethic would have forestalled that agency's overexploitation of the nation's national forests through most of its history. As important as the ethic is, more important is how it was created. What buy-in did it receive during its creation from within the agency it guides? From the broad general public it serves? What steps have been taken to establish it as a fundamental belief of both present and future employees? Ultimately, it is factors such as these that will determine whether the ethic is a decisive force in driving agency practice.

Broad Dialog – Democratize All Conservation Practice

The process used to reach a consensus within a community is as important as the outcome. That is because it is only through an open, democratic process that the true will of a community's populace can be determined. Further, the image of the initiative, how it is perceived by the public, is formed at this time. As with all elements of this initiative, complexity increases with scale. Ergo, some of my comments will only be germane to higher levels such as state and national.

Importantly, it is at this stage that the stakeholder base for the initiative is built. This base, like the foundation of a building, must not only be strong, but must be extensive enough to support everything which is to be built upon it. Trying to add on additional elements at a later date, as with a building, is never as effective. All citizens have a stake in conservation, though many may not be aware of it. After all, clean air, pure water, birds in our backyards, bizarre climate events – we all should care about one or more of these conservation issues. Relatedly, each of these is directly influenced by our individual actions – the actions of every one of us. Ultimately, we must all be part of the stakeholder base. For those who may never have thought about conservation, for whom it may seem an irrelevant part of their lives, this is not due to conservation being irrelevant to their well-being. It is due to a lack of contact with that reality and lack of understanding of how the natural world underpins our whole existence. Just because we may not know where the tap water in our faucets comes from, or the numerous steps taken to keep it clean, does not negate the reality that much is happening outside of our sphere of perception to keep our drinking water available and safe.

Engaging all community members is so important that I have separated it out as framework element #6 which is discussed further in Chapter 9. I raise it here to emphasize its role in the earliest stages of initiative development.

FRAMEWORK ELEMENT #2:
Create a Charter of Community Conservation Values

Dialogue would coalesce in the identification of the community's core conservation values. The process would then culminate in crafting the common threads generated through this dialogue into a simple framework document, perhaps taking the form of a "Conservation Bill of Rights," "Conservation Contract," "Conservation Charter," "Community Conservation Compact" or something of the sort. It must be formulated in a way that would be widely accepted. Those values would provide a benchmark against which policies or actions may be applauded or criticized (Callicott 1994). They would provide an overarching context for conservation actions whether presently underway, or yet to be developed. They would offer a foundation upon which to build more expansive conservation values in the future. And they would represent clear principles to promote among local youth so that such values might be strengthened, rather than diluted, in upcoming generations. Importantly, by being represented in a visible charter, they would serve to create a shared community identity (Sandel 1996).

Martin Luther King famously proclaimed, "I have a dream," not "I have an issue." – VAN JONES

This point could not be better reflected than by retired Air Force officer Mike Penland who commented, "I can remember when the Air Force first put our 'core values' in writing. My commander made everyone in our squadron memorize them and recite them from memory. I thought this was a waste of time. But since then I have realized that those short few phrases are much more than words; they describe what and who we are" (Penland 2013).

Widely-endorsed conservation values will likely be few in number and general in scope. Thus, the charter would be a simple document: short, concise, and to the point. The charter might consist of, say, only six, or four, or two key principles designed to help guide a community towards a sustainable future. That does not matter. What is important are two things: (1) that this process provide a shared vision of conservation values; (2) that the outcome reflects a vision in which the community takes pride and that it is willing to make sacrifices to achieve. This is something which is evident in but few places today. It is something which will create a clear direction for further conservation efforts and enhance social cohesion and commonality of purpose. A common vision of conservation values will enormously expand a community's stakeholder base around this theme as well as strengthen and focus conservation efforts now and into the future.

To reiterate a point made earlier, the values charter need not be conceived of as a legal document. It should be an expression of a broad consensus. I would argue that a consensus on this matter which is genuinely reached will advance conservation further than any formal legal framework could ever achieve.

Many readers may take exception to there being any benefit in producing an informal document, built on a loose consensus, and lacking any legal authority or "teeth" whatsoever. After all, such an approach is totally contrary to much conservation thinking that believes only legislation with teeth has conservation value. To demonstrate the contrary, I offer you the singular case of "WHMSI."

From my first day in the US Fish and Wildlife Service as coordinator for Latin America and the Caribbean, I was charged with the responsibility of implementing the Convention on Nature Protection and Wildlife Preservation in the Western Hemisphere (Western Hemisphere Convention). This was a little-known convention ratified in the early 1940s and virtually forgotten due to the escalation of World War II. The year I joined the Service, we received a $150,000 appropriation which, over the course of a few years, increased to over $1 million to implement this convention. Our problem, though, was that the convention had no governing body or steering committee for its implementation. Having no alternative, we simply executed bilateral grants with other countries to utilize the funds Congress gave us. At the same time, there was always an interest and awareness that the convention could be a more productive tool.

Fast forward 20 years. A meeting was held with our sister agencies from all of the Western Hemisphere's 33 countries to discuss whether there was interest in collaborating on wildlife conservation beyond existing initiatives and, if so, what such collaboration might look like. Non-governmental organizations were invited as observers. Following several meetings, the outcome was impressive – creation of the Western Hemisphere Migratory Species Initiative (WHMSI). With regard to wildlife conservation, the hemisphere had never seen anything like it. Here are its qualities and their impacts:

- An agreement was reached "in principle" by representatives of all 33 nations to collaborate on conservation of the hemisphere's terrestrial migratory species. This was soon expanded to include migratory marine species as well.
- By agreeing "in principle" and not formally adopting or signing a document, WHMSI was not binding and so the foreign

ministries of each nation, purveyors of huge bureaucratic baggage, were removed from the equation.

• The WHMSI agreement had no formal membership. Players could come or go as they chose. But everyone was coming and no one was leaving.

• Agreement in principle enabled countries that could not "officially" recognize one another, such as the US and Cuba, to actually communicate concerning conservation. It enabled all 33 counties to collaborate in a way no other forum could.

• A steering committee was created with representatives of 6 governments, 10 civic organizations, and any interested international body – an unheard of sharing of leadership.

• Under typical international agreements, countries never cede their power to civic organizations, but because WHMSI decisions were by "consensus," any government could halt, without a vote, any projects with which it did not agree.

• Formal conventions became a part of WHMSI due to the expansive forum it provided them, thus the Convention on Migratory Species (of which the US was not a party), the Convention on Wetlands of International Importance, and the Organization of American States (OAS) became very actively involved. The OAS contributed $100,000s to WHMSI initiatives.

• Non-governmental organizations had unheard of access to high-level governmental representatives leading to the former's increased engagement.

• The USFWS was able to implement a key mandate under the Western Hemisphere Convention far more effectively than it could have under any existing mechanism.

• At the 11th Conference of Parties of the Convention of Migratory Species in November 2014, in Quito, Ecuador, WHMSI's Western Hemisphere Migratory Flyways Plan was adopted by that convention as the Flyways of the Americas Plan.

WHMSI's Unique Qualities:

- Agreement "in principle" rather than formally enabled greater participation and reduced bureaucracy
- Informal membership facilitated engagement by distrustful participants
- Enabled communication between nations that did not officially recognize one another
- Decision-making by "consensus" facilitated the sharing of power
- WHMSI could expand its mandate with minimal difficulty
- Because of its diverse representation, its voice and influence grew

There you have the power of an informal charter when built upon goodwill of the participants. While such agreements have certain shortcomings, they have great advantages.

Once identified and reflected in a charter, core conservation values should then be promoted comprehensively so as to become a widely recognized and integral part of the community's discourse, thought, decision-making, and actions.

Hopefully, with time, communities would choose to celebrate their charters on a periodic basis. Just as migratory bird day activities and birding festivals have become a rage in recent decades, events scarcely heard of not many years ago, so the same could be done revolving around community conservation charters. An initiative such as Earth Day, an effort founded around environmental ethics, would be an ideal launching pad for such an activity. This preeminent annual event has the potential to serve as a powerful unifying force for moving conservation values forward at all societal levels simultaneously.

I recall sitting in a meeting, probably in the early 1990s, one of many related to bird conservation that I attended at the time, when Russ Greenberg, then director of the Smithsonian Institution's Migratory Bird Center, out of the blue suggested, "Maybe we should create a 'Migratory Bird Day' to draw people's attention to migratory birds?" I looked up. Now that was a novel and interesting idea. It was elegant, easy enough to launch on a small scale, it hadn't been done before, but so what? Despite having virtually no budget, Russ wasted no time in setting up the first event. His idea grew like wildfire and by 2006 "World Migratory Bird Day" was formally adopted by an international treaty as a global event and today Migratory Bird Day is celebrated annually through thousands of events around the world. Pretty impressive.

Conservation values deserve their day just as do migratory birds, endemic birds, shorebirds, along with the many other wonders worth celebrating in nature.

Additional comments

As mentioned at the start of this chapter, the identification of conservation values and their representation in a charter of some sort, should be a goal for implementation at all levels. The smaller and more homogeneous the community, the simpler the process.

Thus far we have identified shared conservation values and framed them into a "conservation charter." Now what?

Much remains to be done. How do we promote these values? On what groups do we focus? Who carries this out? How might this initiative benefit conservation? These issues and others are the subject of subsequent chapters.

CHAPTER 7

FROM VALUES TO IMPLEMENTATION

Identifying the conservation values of a civic organization, a town, a state, or a nation, and memorializing them in a charter, are important achievements. They create positive aspirations and a shared identity around which the community can rally and better focus its future. In and of themselves, these accomplishments have the potential to empower organizations and institutions supportive of a sustainable environment to refocus their efforts in a manner more in tune with public sentiment. Nevertheless, they are only preliminary steps and, were we to stop here, we would still fall far short of where we need to be if we are to halt and ultimately reverse the negative environmental trends facing us, ranging from species loss to climate change impacts. The identification and chartering of a community's conservation values, whether a town or a nation, provides clear objectives, but much more is required if we are to have any hope of reaching them. This chapter outlines the next steps, as represented by framework elements #3 (frame, implement, and monitor conservation values), #4 (focus on stakeholder groups), and #5 (customize initiatives for each stakeholder group).

FRAMEWORK ELEMENT #3:
Frame, Implement, and Monitor Conservation Values

This third element of the framework introduced in chapter 5 encompasses three components: (1) Framing of values in order to promulgate them in a consistent, compelling, and recognizable manner; (2) Implementing conservation values through specific actions and incorporating them into community processes; and (3) monitoring them using measurable indicators. Let's look at each of these, keeping in mind that the strategies discussed can be adapted and implemented at various levels, thus making them as valuable for a local farmer's association as for a national initiative.

Framework Element #3:

- Frame Conservation Values
- Implement Conservation Values through Specific Actions and Incorporate them into Community Processes
- Monitor Conservation Values Using Measurable Indicators

Frame Conservation Values

The first major component of this framework element – framing of conservation values in a way that is consistent, compelling, and recognizable – is of particular importance in developing greater interest in their acceptance and promulgation (Marvier and Wong 2015). To achieve this in an expert manner, the most up-to-date marketing techniques should be applied to this task by professionals in that field.

It might seem that everything associated with conservation values should be pure and unblemished. After all, they represent an important part of the philosophy by which we live. How could we be so crass as to apply commercial, materialistic marketing techniques to something so special? To speak of "framing" or "branding" our values may seem quite demeaning and inappropriate. Perhaps that is true. But it is also true that following the "tried and true" approach to conservation delivery – each entity developing and promoting its messaging in isolation – is a recipe for failure (Nickerson 2003). All we have to do is open our eyes to the wide array of popular consumables that surround us with which we identify, and we can find behind each of them a well-crafted packaging and marketing strategy to ingratiate them in the public's eye.

Framing and marketing of conservation values in an effective and consistent manner, taking a page from the remarkable success of American business, is essential to maximize effective messaging to the public. We have a wonderful "product" to put on the market – why disdain from promoting it so that every citizen might make conservation values a greater part of who they are?

Do you drink Coca-Cola? Do you drink Pepsi Cola? Do you drink generic cola? Can you tell the difference between them? Quite a few people think they can. Generally, cola drinkers appear wedded to one name brand or another and willing to argue over the difference. Off hand, I do not recall ever meeting a cola drinker wedded to generic cola. Is generic cola really that bad?

The dominance of Pepsi and Coke over other colas is not entirely about taste. It's mostly about advertising. It's about packaging. It's about branding. It's about decades-long marketing. Numbers tell the story regarding how determined Coke and Pepsi are to convince you that they have the best cola on the market – and it works!

How much do you think Coca-Cola spends on convincing you its product is the best? A few million dollars? A hundred million?? A billion??? No, none of the above. In 2019, alone Coca-Cola spent $4.2 billion to sway you towards its products, primarily Coca-Cola. Pepsi spends a bit less.

To give you a sense of the scale of this advertising, the US Fish and Wildlife Service, the primary federal agency responsible for conserving our nation's fauna and flora, has approximately 9,000 employees, at least 567 national wildlife refuges encompassing over 95 million acres of land, and field offices across the country. All costs of this agency could be funded for nearly two full years using Coca-Cola's annual advertising budget alone!

It so happens that in blind taste tests participants frequently change their preference once their blindfolds are off and they see the packaging of the samples, and usually they do so without recalling that they had made a change (Zaltman 2003). This is to say that once participants can see what they drank, they have no qualms about reversing their choice to coincide with their prior beliefs. Effective packaging, branding, and marketing of products have been demonstrated definitively to be by far the best means of increasing sales – to get people to recognize and purchase your product.

"The world has achieved brilliance without conscience. Ours is a world of nuclear giants and ethical infants." - OMAR BRADLEY

Because of the competitive advantages it provides, a well-known brand, whether for selling a product or promoting a cause, is regarded as one of the most valuable assets an entity could have (Levin 2002). One example: A few years ago a dramatic article was published in a popular magazine documenting the appalling slaughter of African elephants for the illegal ivory trade – a massacre which worsens on a daily basis. The article advised readers, in closing, that should they desire to help address the problem, funds could be sent to either of two elephant conservation programs, the addresses of which were provided. One was an initiative sponsored by the journal. The other was the African Elephant Conservation Fund of the US Fish and Wildlife Service, a program administered by my former office. As a consequence, our program was pleased to receive checks totaling over $2,000. The most stirring contribution was from a woman on the island of Guam, who contributed $100 a month from her paycheck to help save these incredible animals. On the other hand, the journal received in the order of $200,000 in contributions.

To what was this ten-fold discrepancy due? Both elephant conservation programs had received equal billing. Individuals who donated were unlikely to have known much about either program considering both are implemented in remote parts of Africa. The reason for the difference was almost certainly the power of "brand recognition." The journal, National Geographic, is widely known and recognized. It is a strongly promoted brand. Conversely, the USFWS program, run by the federal government, is virtually an invisible brand. It is illegal for federal agencies to advertise. The astronomical difference in funding contributions was the direct result.

Branding achieves more than just product sales. Among other things, effective branding creates shared consciousness. This strengthens bonds not only to the product, but also to other individuals who embrace the same brand (Hickman and Ward 2007). Colas aside, branding has also been shown to create a shared sense of moral responsibility (Hickman and Ward 2007). Such is exactly what we strove for in framework element #1 of this initiative in seeking shared conservation values. Branding is a major tool that can reinforce that objective in how we word and promote the charter created in framework element #2.

If branding of a soft drink can be so extraordinarily successful, imagine what can be achieved through the wise branding of conservation values? Because of its allure, effective conservation branding might not be inordinately expensive. As an example, during my tenure in the US Fish and Wildlife Service, Congress legislated that the US Postal Service create endangered species postage stamps, part of the proceeds from which supported our endangered species conservation programs for elephants, tigers, gorillas, and other endangered species. It happened that John Osborn, president and CEO of BBDO-New York, an advertising firm highly acclaimed for the quality of its creative work, found the goals of these conservation programs to be so meaningful and important, that he provided tens of thousands of dollars-worth of pro bono advertising advice and creativity to bolster their success. Noble goals inspire noble actions.

Why do we not apply such an effective tool as branding in broad measure across the field of conservation? In this regard, I am not referring to branding of the conservation groups themselves, many do a fine job at that. I refer to messaging around their shared conservation values. The answer is simple, they cannot claim as much credit for their achievements using that approach – a major concern in generating donations. Not only do these groups rarely coordinate branding of their conservation programs, too often they don't even coordinate their field efforts.

Take migratory bird conservation as an example. Some of the major initiatives aimed at their conservation include: the Neotropical Migratory Bird Conservation Act (NMBCA), Partners in Flight (PIF), Wings Across the Americas (WAA), Important Bird Areas (IBAs), Joint Ventures (JVs), North American Waterfowl Management Plan (NAWMP), Western Hemisphere Shorebird Reserve Network (WHSRN), the Convention on Migratory Species (CMS), and the North American Bird Conservation Initiative (NABCI), among others. I have been engaged in some way with most of these, all of which possess the common thread of conserving migratory birds both in the US and throughout the Western Hemisphere. At best, they associate with one another on an infrequent basis and, to this date, to my knowledge, have never discussed the potential to develop messaging of shared themes. As a specific example, in preparation for a major meeting of the Western Hemisphere Migratory Species Initiative (WHMSI), many of these groups were offered

$10,000 in matching funds to synchronize a gathering of their own memberships as a complement to the WHMSI meeting. The intent of this offer was to broaden the engagement of like-minded groups in discussions about shared interests. Only the Convention on Migratory Species chose to host a simultaneous meeting and benefit from the matching funds, as well as the other advantages of meeting jointly with a partner organization.

"He who masters the power formed by a group of people working together has within his grasp one of the greatest powers known to man."

– IDOWU KOYENIKAN

There are a number of reasons branding jointly by organizations is not applied broadly to conservation projects. We've mentioned that organizations in the non-governmental sector often compete for the same sources of funding and so each group wants to stand out by featuring its own initiatives. Another reason touched on earlier, is that some perceive conservation as a higher calling for which branding is not appropriate. This objection takes various forms: "Conservation is scientifically based and so should not be contaminated by marketing." "Nature is not a commodity." "It is beneath our profession." "We are not a business." Another reason is that conservation professionals are not trained in those skills nor to recognize their importance. Resultantly, most in the profession do not think of conservation in that way.

I do not mean to suggest that such efforts are never attempted. It is simply that such initiatives are the exception, rather than the rule. Also, they are often elaborated by individuals who lack professional training and experience in marketing which results in amateurish products.

As Malcolm Gladwell points out in his bestseller, *Tipping Point*, "There is a simple way to package information that, under the right circumstances, can make it irresistible. All you have to do is find it" (2002). Of course, finding it is most effectively achieved by experts, rather than amateurs.

It is time we learn from the private sector by applying to conservation a tried-and-true product of American enterprise – marketing. Furthermore, we should apply it not following the classic approach of those in our profession – assigning the task to a staff person with little or no experience or expertise in this field. It is time we approach this endeavor with the professionalism and rigor it deserves. It is time we bring on board top-flight experts to craft our messaging in the most effective manner possible.

Despite the fact that our discussion here has focused on reaching out to people, you will notice that the term "environmental education" has not been mentioned. That is because the outreach needed is not environmental education as commonly understood – of simply building awareness (Saylan et al. 2011; Bekoff 2013). What is needed is a much more precise, finely-tuned type of outreach.

Every community's values and culture are changing more rapidly than any time in history, particularly with regard to our increased desires as consumers. This shift, clearly unsustainable, leads to increased environmental degradation which we touched on in an earlier chapter. Marketing has been the principal driver of this change (Dee 2004). The conservation community has the option of standing by and watching our appetite for consumerism consume us, or it can join the fray and market values and cultural change that will lead us to a better future. I suggest we opt for the latter.

Implement Conservation Values Through Specific Actions and Incorporate them into Community Processes

The second component of framework element #3, expressing

our conservation values as specific actions that communities implement by incorporating them into community processes, is another fundamentally important issue that needs special attention and input from experts. An example might be where a community decides not to trim vegetation along town road borders during spring and summer to reduce the killing of breeding birds and mammals. This action would then be incorporated into community processes.

These actions should preferably be simple and of frequent application. They should be specific actions that are to the benefit of "we," the community, rather than "me," the individual. Unquestionably, numerous innovative community-enhancing ideas are long underway. It is a matter of sharing these ideas and spreading them widely so that they ultimately become accepted as commitments that we are proud to meet as part of a community.

Converting general values into specific community actions, individual behavior change, and ultimately the establishment of new norms, like most change, is complicated, but the matter is thoroughly addressed in other works, so I shall not elaborate on it here. Useful resources are books such as *Influencer: The Power to Change Anything* (Patterson et al. 2008), *Tools of Engagement: A Toolkit for Engaging People in Conservation* (Braus 2011), and *Navigating Environmental Attitudes* (Heberlein 2012). The extent to which a community's conservation values are inserted into community decision-making and actions will determine their overall benefit.

Monitor Conservation Values Using Measurable Indicators

Conservation values are general statements, thus the challenge for every community will be to express them in some way that is a **measurable indicator.** This challenge represents the third component of framework element #3 – the need to establish monitoring protocols using measurable indicators. A conservation value might read something like, "We, as a community, take responsibility to leave the earth in good shape for future generations." While that is a very positive statement, how might a community measure success towards its achievement? To that end, it will be necessary to create measurable indicators. Such indicators must be concrete. One example might be agreeing that a given percentage of community land will be kept in green space. Or, that no invasive trees or shrubs will be planted on town property, with a list being made of such species. Developing satisfactory indicators within a community that enable implementation of the values, though a significant challenge, will result in concrete actions that dramatically benefit the community. As with virtually every framework element, engaging professionals with expertise in technical areas such as this will facilitate addressing them effectively.

As communities think about "we" in terms of actions it will take as a body, it would prove worthwhile to consider the related issue of each member's duties to the community. As the great sage Mahatma Gandhi stated, "There can be no rights without duties." If we desire the benefits provided by our communities, shouldn't we also want to contribute to making it better? Improving our communities, whether town, state, nation, or the world, should not be perceived as a chore or an impingement on our freedom. It should be something we take great satisfaction in doing.

FRAMEWORK ELEMENT #4:
Focus on Stakeholder Groups

The fourth framework element is central to promulgating conservation values and entails engagement of each and every community sector. This is dramatically different than how most conservation efforts are conceived. Most begin by the prioritization of species needs. They ask: "Which species are most endangered and under greatest threat?" We begin by prioritizing sub-groups within the community. We ask: "How do we best engage each sector of our community to build its concern for the natural world?" This might be achieved in any of several ways and will vary depending on a community's size. Of singular importance, particularly for larger communities, would be to address **all sectors within the community.** This framework element should be a major focus of local, regional, and national conservation organizations.

Larger communities such as towns, states, and nations, can be looked at as being composed of sectors. Frequently I refer to these as **stakeholder groups.** By "stakeholder" I mean a collection of individuals or groups drawn together by similar interests – farmers, ranchers, teachers, clergy, conservation organizations. Because each stakeholder group is impacted in some way by the initiative's outcome, it should be looked at very much as the initiative's owners.

Stakeholders groups may be divided even further into sub-groups. While stakeholder groups might be "inner-city inhabitants" and "farmers." Sub-groups would be "inner city high-school students" and "corn farmers." From the perspective of our discussion, focusing on sub-groups assists with: (1) ensuring their comprehensive engagement in the initiative; (2) understanding and being sensitive to cultural differences; (3) identification of particular conservation values of special interest to that sector; (4) determining internal leadership within the group; and (5) more focused dialogue, follow-up, and outreach regarding the initiative at a later stage. Looking at sub-groups is valuable from a social sciences perspective because it helps fine-tune understanding of group values and behaviors.

Engaging larger communities ought to include the development of specific **sub-strategies or stakeholder-strategies to reach**

each sub-group. These customized blueprints would combine communication, outreach, and education components. As discussed earlier, to effectively engage all sectors of a community in the identification of conservation values, it is important to consider the common interests and particular concerns of each target audience or stakeholder group. Also important is that leadership for efforts associated with each sector come from within. The challenge is to develop approaches for effectively reaching each stakeholder group so as to better inform and engage them regarding conservation values and to grow their thinking in this regard. It is through the effective creation and implementation of these focused stakeholder-strategies that a great step forward can be taken towards a sustainable future. This component of the framework should entail consideration of new and more effective mechanisms to engage each stakeholder group with regard to conservation values. The groups themselves, supported by the broader conservation community, should take on this task. Outreach programs such as the pride campaigns of Rare, the mass media techniques of WildAid, and the radio soap operas of PCI Media Impact offer innovative approaches for reaching stakeholders.

RARE: https://rare.org
PCI MEDIA IMPACT: https://www.pcimedia.org
WILDAID: https://wildaid.org

Sub-strategies for stakeholder engagement would address many elements. Some are summarized in the box below:

Sub-strategies for Stakeholder Engagement Would Address:

- Communicating the identified conservation values to the group
- Creating mechanisms for their dissemination and implementation
- Integrating this effort into existing initiatives
- Coordinating and networking of on-going efforts
- Developing of new tools
- Designing and implementing a social marketing initiative
- Identifying desirable actions and behavior changes for each value
- Creating of monitoring protocols
- Measuring of success

Sub-strategy formulation should engage a broad cross-section of its particular stakeholder group. Plan development should be as inclusive as possible to ensure that this process, like the earlier one of identifying conservation values, has ownership and support from the constituency. If this is achieved, then there is real hope of a transformation which moves towards a more sustainable conservation future.

FRAMEWORK ELEMENT #5:
Customize Cornerstone Initiatives for Each Stakeholder Group

The fifth element of the framework is the development of what I refer to as **cornerstone initiatives.** While, arguably, these might serve as a component of framework element #4, I have separated them due to their singular importance. Certainly there are initiatives already in place which could be classified as cornerstone initiatives. The Blackstone Challenge mentioned in the previous chapter and the pride campaign to save the Saint Lucia parrot described in Chapter 1 are two examples. Such initiatives should receive special focus. A cornerstone initiative is one specifically crafted to address some basic need or behavior of a stakeholder sector related to conservation values. This need or behavior could be any of many things. Such initiatives should also address underlying causes of on-going, long-term issues facing a sector. Creating them can be challenging, but they have the potential to bring conservation to an entirely new level. For that reason, they are one of the most exciting aspects of the conservation formula discussed here.

The development of cornerstone initiatives offers a number of benefits. In addition to addressing a fundamental need or behavior of the target sector, the cornerstone initiative can serve as a direct, on-the-ground reflection of the overall values initiative and its purpose. A good cornerstone initiative would also get stakeholders thinking about effective, comprehensive solutions to other underlying problems facing conservation. Because of the great capacity of cornerstone initiatives to empower specific stakeholder sectors, it is important that we explore them in a bit of detail.

Whoa! Cornerstone initiatives for specific sectors of a community??? What about tailor-made strategies for the species we need to conserve? Why not cornerstone initiatives for threatened habitats??

Well, that's the biggest difference between the framework we are exploring and most existing ones. It focuses on people who, in turn, we must count on to conserve species. Let's look at some examples of cornerstone initiatives to understand this better.

Making a Real Difference

A few years back, our international program in the US Fish and Wildlife Service had the opportunity to undertake a new initiative with Mexico. The first question was: "An initiative on what? Endangered species? Shared migratory birds? Habitats of mutual interest?" No, none of these. We decided to focus on the stakeholder groups in Mexico that had a substantial impact on species and their habitats, but were not adequately engaged by Mexico's conservation community.

And just how do you determine that? We held a three-day discussion with approximately 50 of Mexico's top conservation organizations focused specifically on that question. The result: A prioritized list of stakeholder groups inadequately addressed by existing initiatives. At the top of the list were **high-level governmental decision-makers** who were **not** environmental ministers. These included national legislators along with ministers of agencies such as commerce, health, and education, among others.

This stakeholder group was intriguing due to its combination of inordinate power, indirect but clear connection with the environment, and its inaccessibility. Due to the latter factor, it was no wonder Mexico's environmental community had not managed to engage this sector. That posed the question: "How could we?"

A steering committee was created to take on the project, a key contributor being a former minister himself. After nearly two years of planning and agreeing to focus on legislators, not agency heads, it was decided that the most effective way to attract and retain their engagement for any period of time was to put them on a boat. Great idea! Once on a boat they could not leave. The boat would cruise the Sea of Cortez for five days, visiting environmental projects and discussing resource-related issues along the way. One important stop would be Cabo Pulmo, where a major industrial complex was in the early stages of creation.

Recruitment of legislators for the cruise was underway when the steering committee received a message – a request from Mexico's president to meet with the committee. During the meeting, the president demonstrated interest in the itinerary, particularly visitation to Cabo Pulmo. He wanted to know, among other things, whether his government had already authorized the permits needed to initiate the massive development. The steering committee responded in the affirmative.

Over the course of the next few months, though the steering committee received no feedback, the Mexican government solicited assessments of the Cabo Pulmo project by reputable entities. Shortly thereafter, the permits were revoked and the Cabo Pulmo project cancelled. No one would have ever considered possible such a positive outcome.

The Cabo Pulmo story reflects a historic success, one produced by focusing on a stakeholder group, rather than following a traditional species orientation. Impressively, it achieved a decisive outcome even before conducting its first activity. This is the kind of impact initiatives can have when they address key stakeholders in the proper way.

The same workshop in Mexico that led to the decision-maker cornerstone initiative and its Cabo Pulmo success, also identified another highly important stakeholder group – Mexico's **protected areas managers and personnel.**

The importance of this group was obvious. At the time, Mexico had over 100 protected areas of various categories spanning tens of millions of acres. It was also generating new ones at an unimaginable rate. Unlike the National Park System or the National Wildlife Refuges System in the US, the vast majority of land in Mexico's protected area system is owned and inhabited by local campesinos and First Nation peoples. As a consequence, these areas experience heavy pressure for humans use, much of which is subsistence. Also, Mexico's sites have great cultural value along with biodiversity uniqueness. These factors combined make protected area management in Mexico a much more complicated practice than in the United States. Being that Mexico's protected area managers and staffs, approximately 600 individuals, were low paid and poorly trained, it was no surprise they were identified as a priority stakeholder group.

The challenge here was, "What might an initiative look like?" "How could we reach these very busy individuals in any substantive way?" We could offer month-long workshops focused on basic management skills, but would these really do the trick? From our prior experience with such workshops, we felt they only scratched the surface.

Engaging Mexico's protected areas managers was far less complicated than was the case for the decision-makers. We had a close association with the head of Mexico's protected area system and so we suggested the following: Given that lengthy training Mexico's present managers was impractical, we would offer partial, but significant support to train Mexico's **future** protected areas managers **if** Mexico agreed that all future hires would have to complete a nine-month training program to prepare them for the job. Mexico would prepare the course curriculum and provide the instructors.

Mexico readily agreed. It immediately coalesced a renowned cadre of personnel from within its agency, universities, and non-governmental groups to develop a curriculum. Not surprisingly, with such extensive input, this took longer than expected. What was a surprise, however, was the outcome.

Mexico developed a 13-month program – four months longer than we had hoped for. The initiative's steering committee felt that was the minimum time necessary for adequate training. Beyond that, not only would all future managers be required to take the training, but all of Mexico's **present** managers would be required to do so as well! Mexico staggered the different training components so that present-day managers could integrate the courses into their work schedules. The outcome was a training program unmatched anywhere in the Western Hemisphere – a program with a major focus on people and communities – a product which far exceeded anyone's initial expectations. That is the power of a solid cornerstone initiative.

A fine cornerstone initiative in the US is that of a group of ranchers in Southeast Arizona and Southwest New Mexico known as the Malpai Borderlands Group. As such, they collaborate to sustain wild landscapes as a part of their way of life. This organization has thus far conserved 78,000 acres of private land through conservation easements with the aim of restoring ecological diversity and productivity and ultimately maintaining this acreage as natural wildlife habitat and productive ranch lands. Importantly, similar groups of ranchers such as the Quivira Coalition are forming across the west, and efforts are underway to facilitate communication and cooperation among these rancher-based initiatives. Looking at what might be considered cornerstone initiative in the inner-city, for over twenty years the city of Chicago has developed innovative initiatives that have made it a leading environmentally friendly city. This includes planting more than a half million trees, mandating the construction of buildings which are environmentally friendly, and installing rooftop gardens on city-owned buildings. Other efforts include creation of a green taxi program that provides incentives for alternatively fueled taxicabs, development of farmer's markets, and establishment of bike-to-work days. Chicago Wilderness is a diverse regional alliance dedicated to just this sort of transformation towards living in greater harmony with nature.

Once cornerstone initiatives are piloted and refined, they then should be replicated so that increasing numbers of stakeholders might reap their benefits. Certainly, more than one such initiative could be developed per stakeholder group, but one or a few high quality and well-focused initiatives will have more impact than a proliferation of numerous, weakly focused, and poorly executed efforts.

It is important to recognize that several groups should be of especially high priority in conservation outreach efforts and regarding cornerstone initiatives. Among these is youth. Another is the conservation professional, for which the framework of a cornerstone initiative is presented in the next chapter.

CHAPTER 8

ONE CORNERSTONE INITIATIVE: TRAINING CONSERVATION PROFESSIONALS

Customizing cornerstone initiatives for each community sector – the aim of framework element #5 – is of special importance yet, at the same time, it can easily be misinterpreted. So that we might better understand this concept, we will explore a specific example of what such an initiative might look like to see the power it brings to promoting positive change.

To that end, I shall lay out a potential frame of reference for restructuring graduate training of future conservation professionals as an example of what such a cornerstone initiative might look like. But first, it is important that this cornerstone initiative be placed in context. So, bear with me as I first provide some background and relay a few powerful incidents that help provide perspective.

No sector would be influenced more by the changes suggested in this book than the conservation professional, and no sector is more central to this initiative's success. By "conservation professional," I refer primarily to those individuals who undertake graduate training in wildlife biology, wildlife management, natural resources management, conservation biology, or related degrees, and subsequently fill professional positions. One of the major challenges in reorienting our approach to conservation would be in the training of future conservation professionals so as to empower them with a new suite of skills necessary to deliver effectively this new approach.

Change will be challenging, but essential for existing training programs to transform themselves to emphasize the importance of working with communities and their values, coupled with the various other reforms necessary to prepare future conservation professionals to be effective in the 21st century. Stated bluntly by Infield and Mugisha (2010) "It is not easy to mainstream cultural approaches within the existing conservation infrastructure due to strong organizational cultures and traditions and institutional inertia. Organizations will have to change their ways of thinking and working. In particular, the privileging of the values of educated elites over those of uneducated and often marginalized sectors of society will need to stop." Regrettably, this statement remains all too true today suggesting that serious change in this field may have to be spurred by outside forces and certainly will require outside input.

Perspectives on Training of Conservation Professionals

Over 80 years ago, the iconic conservationist Aldo Leopold identified the magnitude of the necessary change, though his insightful words have gone relatively unheeded: "Our profession began with the job of producing something to shoot. However important this may seem to us, it is not very important to the emancipated moderns who no longer feel soil between their toes. We find that we cannot produce much to shoot until the landowner changes his ways of using land, and he in turn cannot change his ways until his teachers, bankers, customers, editors, governors, and trespassers change their ideas about what land is for. To change ideas about what land is for is to change ideas about what anything is for. Thus we started to move a straw, and end up with the job of moving a mountain (1940)." As suggested in this quote, "producing something to shoot" was the focus of wildlife management through much of the 20th century, a situation we discussed early in the book. Though wildlife conservation has shifted moderately since that time, the essence of Leopold's remarks remain insightful – conservation requires the engagement of many more societal sectors than most conservation professionals tend to realize. It is that broader constituency to which Leopold is referring when commenting on the "job of moving a mountain." The conservation community, to a large extent, has chosen the easy task – keeping conservation's traditional proponents happy. This book suggests not only that we take up moving the mountain, but also how we might go about doing just that.

The change in training of conservation professionals Leopold called for, the need to move a mountain rather than a straw, is even more necessary today as a result of our population's dramatic shift off the land and into urban areas. Despite this circumstance, all too many conservation professionals fail to adapt to Leopold's vision because of how they themselves think and interact. The renowned zoologist George Schaller, late in his career, recognized the same point Leopold expressed and lamented, "I have in recent years focused less on detailed science, something I enjoy most, and more on conservation. I have tried to become a combination of educator, diplomat, anthropologist, and naturalist – an ecological missionary, balancing knowledge and action" (2011). Kennedy and Thomas (1995), among others, call for major changes towards the training of future conservation professionals in skills necessary to address societal values, but in my mind they do not go far enough.

Major change in the training of conservation professionals on a broad scale is long overdue. Ultimately, that change should result in a reshaping of the conservation profession and, as Peter Senge, a leading expert on organizational change, noted, "new jobs with new skill requirements … in turn demand new kinds of people" (1994).

Our Inadequacies at Saving the Tiger Reflect Failed Training of Top-level Professionals

The change that I believe is needed is not confined to conservation professionals operating in the United States but extends worldwide. This was vividly brought home to me in 2011 at a major international conference on tiger conservation in which I participated. India had just completed a major survey of its tiger population, a survey which reputedly involved 100,000 people, as noted in an earlier chapter. The 13 countries in which tigers presently survive – the "range countries" – had all recently completed national tiger conservation action plans, and these had been compiled into a single comprehensive book. To celebrate the survey and explore next steps regarding the national action plans, India hosted a conference of the range countries, as well as all other interested entities, both governmental and non-governmental alike.

The conference was illuminating. Highlighted was that the survey detected approximately 1,700 tigers throughout India, a number very close to that derived from a less comprehensive survey four years earlier. In addition, India's environmental minister made three ominous points at the inauguration of the meeting:

1. India is committed to six percent economic growth per year.
2. India lost ten percent of its tiger habitat during that period.
3. Nearly one-third of India's tigers occur outside of its tiger reserves.

Several questions immediately arise: No matter what the survey reported India's tiger numbers to be, how long can those numbers be sustained if the country is experiencing a ten percent loss of tiger habitat every four years? How can those numbers be sustained in the face of six percent economic growth per year? And how can those numbers be sustained if one-third of the country's tigers wander outside tiger reserves and face an inevitable increase in conflict with people?

Fortunately, the conference brought together what might well be considered the global authorities and practitioners of tiger conservation, thus providing an excellent opportunity to grapple with these difficult issues, issues being experienced in varying degrees by virtually all of the range countries represented. What better than a forum to address fundamental, underlying threats to one of the world's most precious animals?

Unfortunately, not one minute was spent on any of the three ominous issues – unquestionably the three most important ones. This was not for lack of time. No, the participants apparently found it more important to discuss tiger surveying. When would the original survey data become available? How comparable was data collection between the present and past surveys? When would the next survey be undertaken? Comments even got down to a discussion of guidance for future survey participants – how they should refrain from talking too much during surveys so as not to disturb the animals and bias the data. Notably, one of India's most distinguished tiger experts insisted that the most crucial next step was to ensure that the subsequent survey be conducted the following year, rather than four years hence, as proposed by the Indian government, a survey, by the way, that cost $2 million to conduct.

Notably, it was this same expert whom, a few years earlier, sat in my office describing how special his research site was due to its overflowing number of tigers. In response, I inquired about the impact such abundance of this large predator might have on local communities and what he was doing about it. Confidently, he informed me that he had a plan for exactly that, a plan which would be put into action in the next year or two. The following day the expert forlornly informed me that his research facility had just been burned down by residents of the area. Apparently even this harsh event had not taught him one iota about the importance of people in the conservation equation.

This august body I was observing, a group which may be said to hold the survival of the tiger in its hands, chose to ignore the tsunami-scale threats to perhaps the most majestic animal on the planet and instead focus solely on survey details – a matter which has little direct impact on tiger survival. Undoubtedly, it was experiences of this sort, nearly two decades earlier, which led tiger expert George Schaller to lament, "The human dimensions of tiger conservation, the inevitable conflicts, have received little notice" (1995).

Especially disconcerting was that this discussion was unfolding in India, the nation whose remarkable success we have explored with regard to conserving wildlife.

The fundamental driving force that has enabled India to sustain virtually half of the world's surviving wild tigers – its socio-cultural values – went almost totally unrecognized at this gathering. The Indians themselves appeared unaware of it. Now, within India's community of tiger experts, replication of western strategies and training leads the way forward. The westernization of conservation strategy reflects a much broader trend of westernization of Indian society as a whole. As Judi Aubel (2010), executive director of the Grandmother Project which uses cultural values to improve the lives of women and children in India, concisely points out, "Globalization involves a virtually one-way dissemination of western cultural images and values toward non-western societies." Suffice it to say, the single most important tool sustaining tigers in India – that nation's values – are shifting before our eyes, and most tiger conservation professionals do not recognize the dire consequences for tigers and all of India's natural heritage, which will derive from this shift. As great a threat as economic growth and the reduction of tiger habitat may be to India's tigers, I would argue that a greater one is the shifting of India's socio-cultural values away from its historic base towards a Western World perspective.

There was one notable instance during the meeting when an American took the floor and suggested that the group's focus was decidedly off-base and it was India's distinctive culture that should be looked to for solutions to the tiger crisis. There was no further discussion of this point but, upon adjournment, a Nepalese delegate approached the speaker and said, "It took a white foreigner to come here to tell us what we should have seen ourselves." It is instances such as this, ones that, regrettably, are far from unique, that have prompted me to touch on the neo-colonial and relatively elitist nature of conservation throughout this book.

It would be reassuring to think that some quirk or other drove the focus of the meeting off base. The *Global Tiger Recovery Program*, a report handed out at the conference that detailed the conservation actions contemplated by the tiger range states, suggests that not to be so. If anything, the recovery program is a concrete confirmation of the degree to which the tiger conservation community is out of touch.

I scanned the document for several key words which reflect important conservation focus areas. Reference to **surveying and monitoring** was made thirty-four times. **Capacity building,** a fundamental and much needed tool to improve the skills and capabilities of local staff and institutions in the various nations so that they might conduct their work more effectively, was mentioned only eighteen times. A third key focus area I scanned for was **public awareness and outreach.** Considering that the tiger is the greatest terrestrial predator on earth, its conflict with humans is a rather thorny affair requiring substantial awareness-building and communication of effective community mitigation measures. We might expect this would be a much-referenced theme. But there were only five references to this topic across the thirteen range countries. More alarmingly, there was only one reference to the need for **education.** And finally, worst of all, **there was no mention at all of *cultural values* – the factor singly most responsible for the survival of tigers in the wild.**

Global Tiger Recovery Program References (13 countries)	
• Surveys and monitoring:	34 times
• Capacity building:	18 times
• Public awareness/outreach:	5 times
• Education:	1 time
• Cultural values:	0 times

The Need for Improved Professional Training

Consideration of this example, among others in the book, suggests the urgency of training future professionals differently, the magnitude of the change necessary, and hints about the skills conservation professionals should possess. The initiative that Jack Turnell, my rancher friend, and I developed was focused specifically on the training of conservation professionals and serves as the core of the material presented here.

The reason so much of what conservationists do is off-base is a consequence, I believe, of our training, particularly our formal graduate studies. They are a necessity for entry into most mid- and upper-level employment in the conservation field. Though clearly progress has been made in some graduate program content, it is not enough. Ultimately, without a major transformation in the way conservation professionals are trained, it is unlikely adequate redirection and refocusing of conservation programs and initiatives can be achieved.

Graduate level training is but one component of the professional conservationist sector's strategic needs, but it perhaps is the most important component. This potential cornerstone initiative is presented both as food for thought and to demonstrate the substantial amount of change in graduate training necessary in the conservation field if it is to be relevant in the 21st century.

Graduate school programs: The practical place to begin exploring the realm of graduate training is by answering the question, "What are the most important skills a professional conservationist should possess?" I look at the answer this way: If conservation is fundamentally about people, their values, and how they practice those values, then the training of conservation professionals should be built around providing the skills necessary to work with people, particularly with regard to values, attitudes, and behaviors which influence the natural environment.

It appears evident that the present content of most graduate school programs is quite different, most requiring a core set of science courses from the ecology and natural resources fields, complemented by an array of electives from the social sciences and economics. Perhaps most noticeable, the Cooperative Research Units (Coop Units), established in 1935 to facilitate graduate training in fisheries and wildlife sciences and sustaining 40 graduate programs in 38 states, managed by the US Geological Survey no less, scarcely seems to have adjusted to accommodate a more modern vision. The primary courses in such programs are not those most needed by practicing professionals.

History of development of conservation programs: Why does conservation training pay so little attention to social skills and cultural attitudes, and for how long has this been the case? As suggested in the Introduction and Chapter 2, in the early days of the conservation movement, a focus on applied wildlife science made much more sense. There was more empty space, fewer people, and remarkably little was known about the animals we were trying to manage. Creation of the Coop Units mentioned above is strongly connected to state fish and game agencies, thus they possess a significant hunting tilt. Though conservation needs have changed dramatically over time, such training programs have not adapted adequately along with them.

The study of wildlife management, with the Coop Units at its center, was gradually expanded to include non-game species to a limited extent. But this change was considered inadequate by many. In the 1980s, the phrase "conservation biology" was coined to express the essential relationship between conservation and the science of the natural world (Soule and Wilcox 1980). It focused on all of the natural world equally, not in a manner that favored game animals over non-game species. This created a fundamental dichotomy within applied conservation science and was a challenge to the status quo regarding game animals. Since then, the social sciences have increasingly become a component of both the Conservation Biology and, to a lesser extent, Coop Unit schools of thought. But, in neither case has it been emphasized to the extent necessary.

The field of conservation biology has continually made strides in the right direction. Most graduate programs in this area include coursework outside of the biological sphere, such as sociology, psychology, and anthropology. That is a good thing. On the negative side, the term "conservation biology" itself misrepresents just what conservation is truly about. Conservation is not mainly about the biology of species and their habitats. It is about people. Speaking generally, graduate training in conservation biology, though now recognizing the need that such training be interdisciplinary, maintains that the biological sciences remain the core of such programs and training in other sundry skills merely complements that base. The acknowledgment that there's more to conservation than biology is a step in the right direction, but the rubric alone sustains the misconception that the primary focus of conservation should be the science of animals or the natural world, not working with people. In line with this perspective, it is reassuring to see efforts to create new disciplines, such as conservation social science and conservation psychology.

To the extent that the interdisciplinary field of conservation biology shifts to a focus on people as the drivers of conservation, rather than the biological sciences, it will be all the more effective.

Many professionals in the conservation field will likely argue that it is dangerous to reduce attention to the biological sciences as the driving force for conservation. After all, as Michael Manfredo (2008), a leading human dimensions researcher noted, "The ideal of the wildlife professional has been to emphasize science while striving to exclude emotional considerations from the decision-making process." I am not at all suggesting that the biological sciences be eliminated from the conservation equation. Biological sciences should remain a fundamental tool of conservation. What I am suggesting is that they be put in their proper place as an important but ancillary component. The biological sciences simply cannot remain as conservation's primary tool. Continuation of such an approach is the equivalent of trying to drive a square peg into a round hole. The biological sciences should **advise** conservation, they should not **be** conservation. They are **a** tool, not **the** tool. Wildlife biologists, ecologists, wildlife managers, botanists, zoologists, have much to offer conservation. They should not necessarily lead the effort. By putting the natural sciences in their proper niche as a conservation tool, I believe the lessons which science teaches us about the environment actually have greater potential of being achieved than they do under present circumstances. A social sciences-based focus, with conservation values being a prominent theme, provides a much more practical and effective approach to reaching and engaging conservation stakeholders who, in the end, determine our path forward and to what extent conservation will be a part of it.

Not long ago I was invited to serve on a leadership team of 15-20 managers to suggest how best to use $40 million by the US Fish and Wildlife Service to address climate change. The three general categories to which funds could be designated were "adaptation," "mitigation," and "education." Generally speaking, "adaptation" involved short-term actions to reduce possible impacts, but ignored the underlying problem. "Mitigation" dealt with reducing emissions. "Education" addressed getting people to care about the issue so that they might take increased action to address it. As you might expect, I believed education, in a broad sense, was most important, because implemented properly, it would have the greatest long-term chance of effecting a reduction in climate change. To that end, I urged that at a minimum we should aim for a balance among the categories. The outcome? Over 90 percent of the funds were earmarked for adaptation to existing initiatives, a small amount went to mitigation, and virtually nothing to education. There simply was no recognition by the agency of the need to engage the nation's citizens to address a problem affecting all of us.

A Cornerstone Initiative:
Training of Future Conservation Professionals

What, then, are the skill sets a conservation professional should possess in order to cope adequately in today's world? We begin with recognizing that training should be built around people-skills with a focus on addressing values, attitudes, and behaviors.

Key Skills of the Conservation Professional:

- Developing trust
- Working in teams
- Excelling at verbal communication
- Collaborating in culturally diverse settings.

Developing trust: Probably the most basic skill, more important than all others in this regard, is an ability to develop trust with people. Trust is the most important part of the equation. Without trust, little can be achieved. In the words of learning guru Peter Senge, "the best systemic insights don't get translated into action when people don't trust one another and cannot build genuinely shared aspirations (1994)." And, with regard to the ability to lead, surveys demonstrate that trust is the single most important factor upon which followers evaluate leaders (Chaleff 2009). The renowned TV show hostess, actress, and trend setter Oprah Winfrey emphasized during a nationally televised speech in 2024, "I'm telling you the truth, that values and character matter most of all in leadership and in life ..."

This sounds simple enough. But it is more complicated than it might at first seem. For one, developing trust is not just about being able to talk to and get along with people, though that is a starting point. It requires empathy for their needs, and caring for their interests.

One reason that building trust sounds simple is because most people think that it comes naturally and that they can do this very well. A telling study found that 25 percent of college students surveyed felt they were in the top one percent of individuals with the ability to make friends. That means, based upon this survey, that 24 out of every 25 students had a misperception in this regard.

The critical importance of the development of trust, requiring as it does the capacity to be sensitive, to listen, and to be empathetic, as a central skill of the future conservation professional is well illustrated by a couple of examples.

In the mid-1980s, I served on a US Department of State delegation visiting the Caribbean to negotiate a series of protocols. One of those addressed accidental contaminant spills by ships at sea in the Caribbean. A central issue in developing the protocol was how to define a "contaminant." What was a true contaminant? Was there such a thing as a harmless spill of something like a tank of fresh water into the sea? To that end, discussions were held in Washington, DC to establish a position in advance of meeting with the nations of the Caribbean. After thorough consideration, it was decided that making a list of contaminants to be covered by the protocol did not make sense. Such a list would be lengthy and would fall out of date almost immediately as new products regularly entered into international trade. Our position, then, would be to endorse a broad definition of "contaminants" so that it would cover everything which might be spilled from petroleum to plastic cups and, thus, not require development of a contaminant list.

This made sense. Our position was aimed to be both practical and supportive of the need to assist Caribbean nations during such emergencies. As backup in support of our position, we developed a comprehensive list of reasons why our approach was superior to the alternative – making a list of contaminants.

The Caribbean meeting was fascinating and, indeed, a lesson in diplomacy and trust. The US delegation confidently put forward its suggestion of how to address the contaminant issue. The nations of the Caribbean graciously proposed the opposite approach – they preferred developing a list of contaminants. No problem – shift to Plan B. The head of the US delegation proceeded to present more and more arguments to support our position. The Caribbean nations were unmoved.

Negotiations went on for several days. But the more the US put forward new arguments for its alternative, the more intransigent became the Caribbean. The outcome? No agreement whatsoever.

The irony was that early on in the discussion, it was apparent that both sides desired the same outcome. The Caribbean wanted US assistance with virtually any type of spill. The US was interested in providing exactly that assistance.

So what went wrong? If one listened carefully, there were hints that a number of Caribbean nations did not trust the US to do what was best for the region. Of course, this was not stated outright. To have done so would have been impolite and is not the customary way most Caribbean nations deal with disagreement. On top of that, being direct could possibly generate negative political consequences. No, the simple way to deal with the matter was to take the opposite tack from that of the US – in this case, to insist on a list despite strong arguments to the contrary. Had the US initially proposed development of a contaminants list, I believe the Caribbean nations would have argued for the opposite.

The failing of the US delegation leadership was its determination to "win" the discussion through reason and debate. It simply did not recognize that many other countries put **trust** far ahead of **reason** in the decision-making process. Without first having trust, consensus on issues, even where both parties desire a similar outcome, can be elusive.

The US Department of State failed to recognize the significance of those drivers, particularly skepticism regarding US intentions, on the outcome of negotiations. We failed to recognize the basic importance of trust. The rotation of US Foreign Service personnel assignments every two to three years has benefits in terms of expanding the breadth of experience and knowledge of its employees, but when applied in a rigid manner, it fails to recognize that, as with most human interaction – diplomacy with most countries hinges on personal relationships and rotating of personnel sets this back.

At a minimum, these negotiations should have been preceded by a pleasant "get to know you" gathering, at least by the heads of delegation. A relaxed interchange on non-business topics could have made all the difference. Had trust been established up front, an accord almost certainly would have been reached in no time and all parties would have departed with their goals satisfied. Instead, not only did a lack of trust lead to no protocol being produced, but the fractious negotiations only furthered the divide between the US and the Caribbean nations making future discussions that much more difficult.

Hardly are such misperceptions limited to any one entity. I shall give another example, this one related to trust and the impact a lack of sensitivity might cause.

The North American Bird Conservation Initiative (NABCI) is devoted to the conservation of all species of migratory birds which occur in the US, Mexico, and Canada, with each of the three countries having a national committee responsible for its own domestic implementation. A few years ago the US committee was concerned about what appeared to be a lack of engagement by its counterpart in Mexico. When consulted on the matter, I suggested a joint meeting so that the two groups might better understand one another's perspectives and priorities, thus leading to improved cooperation.

The two national committees met in Mexico City, each nation early on presenting its primary needs and interests regarding migratory birds. Top priorities on the US list were objectives such as obtaining additional data on particular migratory species and setting aside habitat in Mexico for such species. The number one priority on Mexico's list was capacity building. Unlike the US, Mexico does not single out specific migratory birds as important. It looks at fauna and flora conservation more holistically, trying to conserve habitats and ecosystems that are shared by many species. Whether looked at from this holistic perspective or the more species oriented one, Mexico reflected that it was far behind the US concerning the capacity of its personnel to effectively deliver conservation on the ground. Consequently, building the capacity of its professional staff, whether governmental or private, was a must if Mexico was to have any hope of being an effective partner in this continental effort. Now, for the first time, it seemed that key players from both countries were on the same page and true progress could be made on the migratory bird conservation front. Then again, perhaps not. We set about developing strategies to achieve the top priorities on each nation's list.

The lead spokesperson for the US was clever. He proposed that Mexico's top priority, building human capacity, could be addressed by treating it as a "cross-cutting" issue. As such, it needed no special attention. Capacity building would simply be made a component of all projects. While this may have sounded constructive, capacity building had long been categorized in this way and, as a consequence, had received little serious attention for years in US migratory bird conservation efforts. This suggestion, in addition to radically diminishing any chance for Mexico's primary interest to be addressed, resulted in only a few minutes being spent to discuss that nation's most important conservation concern. With that out of the way, the remainder of the meeting, a full day, could be dedicated to US priorities, despite Mexico's clear indication that it preferred a different approach.

By the end of the meeting, the US committee felt triumphant. It believed it had now developed trust and a rapport with the Mexican committee and a plan forward had been adopted. Well, not really. To this day, everyone is still waiting for the plan, developed that day long ago in Mexico, to be implemented. Besides, even if implemented as agreed, the long-term success of such an effort had little chance of being sustainable over time. The fundamental reason why, as Mexico had made clear – that country did not have the technical capacity to address such complicated matters effectively. Beyond that, the Mexicans took away a different impression from that meeting. Their concurrence with the US proposals derived not from true agreement with what should be done, but rather reflected politeness and not wanting to display disagreement with their strong-minded US counterparts. The one-directional "trust" perceived by the US delegation was not reciprocated. The outcome – cooperation on migratory birds reverted to the condition it had been in before the meeting – not to mention it left a bad taste in the mouth of the Mexicans which, if anything, set back collaboration. One-directional trust is the equivalent of no trust at all.

I have presented the marine contaminant protocol and the migratory bird discussions because they are particularly poignant. They occurred at the highest levels, had regrettable consequences, and their outcomes were singularly related to a lack of trust in the first case and insensitivity leading to lack of trust in the second. As long as we continue to give lip service to the importance of building trust, and minimize what that entails, progress towards conservation, or any other goal, shall continue to falter.

An obvious question deriving from this discussion is how, then, do we develop skills relating to trust? How do we train for it? Trust is an interpersonal skill. We do not learn it by going out into the forest and studying some animal or ecological process. Neither the jawbone of a deer, nor the feathers of a bird, teach us much about it. This skill is built and refined by working with others, and by coaching. Even more so, it is built by depending upon others. It is built by being sensitive and listening carefully. Having respect, too, is essential. One must learn to analyze social situations and understand the need for trust in human affairs.

Working in teams: One of the most powerful tools for developing trust is, in fact, another basic skill associated with the training of future conservation professionals. That skill is the ability to work effectively in teams. Teams are of great value in honing interpersonal skills and strengthening trust. Working in teams should be the rule, not the exception. To the extent possible, all learning experiences should be built into the team concept so that, over time, working collaboratively becomes second nature to participants. And, contrarily, working alone becomes uncomfortable.

Teams also provide the benefit of being more effective at addressing the complex conservation issues facing societies today. The world is simply too complicated for lone individuals to tease apart its intricacies – what makes things tick. In other words, understanding the challenges species face to survive, and communities face to live in harmony with the earth, takes the input of many. Every day individual problem-solving, whether related to conservation or some other enterprise, is becoming increasingly obsolete.

Beyond having program participants work in teams with fellow students, it is equally important, if not moreso, that they work in teams with the communities and other entities outside the training program with which they must actively interact. Students, ideally, would become immersed in the practice of developing trust in communities and, conversely, building the trust of communities in them – through participation in teams.

At present, many conservation professionals find working in teams to be inherently threatening, my experience suggests. This is a major reason why potential students in such a graduate conservation program should be recruited for their specific capacity in interpersonal relations and their desire to refine such skills. Clearly, we are looking here for a very different character profile than at present for acceptance into a program. Michael Manfredo, a professor at an innovative program at Colorado State University oriented in this manner, has highlighted the importance of dealing directly with the topics of emotional norms and emotional intelligence (Manfredo 2008). Yes, it is important that selected participants appreciate nature, but this cannot be at the exclusion of respecting and appreciating people. Participants must have hearts big enough for both – animals and people.

The capacities mentioned above can be selected for. Evaluation tools for emotional intelligence assessment provide a sense of an individual's interpersonal skills including empathy, social responsibility, and the capacity to establish relationships. They can also assess a person's adaptability to new situations and capacity for stress management. Such tools are valuable in student selection, as well as to measure on-going development. With constant practice, critical interpersonal skills can be honed, thus enhancing an individual's capacity to work with others. Various elements of an effective training program can build these skills. Exercises and role-playing activities have been developed that help to understand, appreciate, and practice such abilities.

Excelling at verbal communication: Following upon trust and teamwork, the next important skill set is communication. Though there are many forms of communication, and all are important to the conservation professional, I refer here primarily to verbal communication. This is communication with peers, detractors, decision-makers, and people from all walks of life. The more effectively conservation professionals can communicate with others, the more successfully they can build the necessary support to achieve meaningful and sustainable goals. Not to be forgotten in the mix, communication, like trust, must be a two-way street. Effective communication requires listening, sensitivity, and empathy.

"Powerful and sustained change requires constant communication, not only throughout the rollout but after the major elements of the plan are in place. The more kinds of communication employed, the more effective they are." – DEANNE AGUIRRE

I offer one telling story pertaining to communication and how poor we are at it: A cohort of 75 college students, including myself, was posed the following question: "A farmer sells a horse for $10, buys it back for $20, and sells it again for $30. How much money did he make?" Sounds simple enough. This is a mathematical question, so there should be only one answer. I was more than startled, therefore, when four answers were proposed by the students, the group divided nearly evenly among them. Some students thought the farmer had made $10, others thought $20, others $30, and still others suggested some other answer. The students were then divided according to their responses, each group meeting to refine the arguments favoring its position and to select a spokesperson. Following presentations by each group representative, all students were offered the opportunity to change groups. The result? Practically no one budged. Nearly everyone was either convinced they were right, or was too embarrassed to demonstrate in front of everyone else that they were wrong! This is quite a result considering that three-quarters of the participants had the wrong answer. I have conducted this exercise on several subsequent occasions with the same result! The take-home messages? One, we are terrible listeners. We only hear what we want to hear. And, listening is an essential part of communication. Two, if people refuse to accept the truth concerning a simple mathematical transaction, one for which there is a right and wrong answer, and in which they have no vested interest in the outcome, no wonder there is such disfunction when more complex issues are on the table. Three, communication is much more complicated than we widely perceive it to be. We presume we are getting our message across when we are not. All the more reason to include it as a fundamental skills required of future conservation professionals.

Collaborating in culturally diverse settings: The future conservation professional must be comfortable in culturally diverse settings. Since all segments of each community and society as a whole are stakeholders with regard to conservation, conservation professionals must be comfortable among them. Too often, conservationists sing to their own choir. At the same time, conservation success does not ultimately hinge on our relationships with others like ourselves. It is determined by the extent to which we expand the appreciation of nature and wise conservation practices to a much broader and more diverse base than at present.

To facilitate a greater appreciation, sensitivity, and empathy for diverse society and enhance the comfort level of conservation professionals to embrace the challenge it presents, I believe working with different stakeholder communities must be integrated into the training program to the maximum extent possible. Students should experience the breadth of stakeholder groups that make up society, the differences among their perspectives, and the complexity of engaging them as important components of the conservation equation. Working with diverse stakeholder groups provides an opportunity to develop trust, communication, and sensitivity skills that are so important in successful conservation work.

What does working with a particular stakeholder group or community look like? What does it entail? One potential scenario would be to have a team of students collaborate with a stakeholder group to address a particular conservation-related problem with which that group is grappling. With a group of ranchers, it might involve exploring how cattle and prairie dogs might coexist more effectively without having either to poison the prairie dogs or remove the cattle. With farmers, it might be how to manage crop edges or harvesting regimes so as not to diminish yield while at the same time improving the survival rate of breeding grassland birds. In the business sector, a project might entail how a company could refine its practices in order to manage more sustainably some natural resource upon which it depends.

A lengthy list of potential stakeholder projects for a team of students to address is easily developed. Importantly, such a list should be created and prioritized by the stakeholders themselves. This will greatly enhance them taking the initiative seriously and engaging the students.

The duration of each stakeholder based problem-solving exercise would depend upon the magnitude of the problem being addressed and the constraints of the training program.

Field projects of the type described here are becoming a more frequent component of university training. What is less the case is the focus on the importance of student success being based upon development of the skills described above – trust, teamwork, communication, sensitivity – rather than the desired conservation goal. Stated another way, though the conservation goal of a field exercise might be the development of a modified grain harvesting strategy within a farming community where bobolinks breed, for example, I believe true success would best be measured not by the plan which derives from the exercise and its completion, as much as from the interpersonal skills students develop in the process. Much preparation and careful planning must go into each exercise, plus the students should receive regular coaching and feedback from their instructors.

Early in the program, students would develop their appreciation and skills working in teams with local stakeholder groups. In later stages, their efforts would focus at a higher level, perhaps even state or federal, to better understand the nuances of politics and decision-making in its various forms. This could be achieved most effectively by exposing students to these processes first-hand. During the final semester of training, they might work directly with legislative bodies, permit granting agencies, or other entities with legal jurisdictions. If such experience can be provided through a joint problem-solving exercise, so much the better. Internships would be an excellent second choice.

Other Basic Skills of the Conservation Professional:

- Attitudinal and behavior change and other basic social sciences
- Problem-solving and conflict resolution
- Group mobilization and organization
- Understanding role of values in conservation
- Basic ecology and environmental sciences
- Social marketing

Other basic skills of the conservation professional: There are various other skills valuable to the future conservation professional. It would be important that students receive basic training in problem-solving and conflict resolution as well as guidance on how to mobilize and organize groups. Obviously a portion of graduate study needs to be dedicated to understanding the role of values in conservation. Such studies would include not only becoming intimately familiar with the articulation of values of the community in which the student might work, but also with the science and nuances of attitudinal and behavior change, and the basics of various social sciences. The same is the case for learning the fundamentals of social marketing. Some basic ecology and environmental science would be fundamental, as well as learning to look at society and the environment in a holistic manner. Training must shift from focusing on species to ecosystems, from refuges and protected areas to watersheds and broad swaths of land, from collaborating only with those who think like us to partnering on a markedly broader scale and seeking out those with different views, and from introspecting about "my" project to sharing about "our" project.

A program as described here would be highly interdisciplinary. It would be built around experiential learning. The expression "learning by doing" has become a cliché, but this makes its practice no less important. The more that experiential learning is used as a training tool, the more effective the training initiative will be when it comes to putting values-based conservation into practice. A program to train future conservation professionals will only be as successful as the breadth of experience it provides. It is because we learn by doing that the practice of teamwork must be just that – practiced.

In line with the changes suggested, consideration should also be given to modifying how student performance is measured. Since training shifts more toward working with people on behalf of conservation, measures of performance should shift accordingly. Too often the measurement of successful student program completion is via a tool totally inappropriate for that purpose. The writing of term papers or a thesis are excellent examples of this. While term papers and theses are tangible products relatively easily graded, all too often they are used as an easy way of complying with academic standards and portraying the appearance of academic rigor.

The tools for measuring student success should stay focused on the skill sets desired of all graduates. Since building trust, teamwork, being an excellent verbal communicator, and having the capacity to work comfortably in socially diverse settings are among the most important skills expected of graduates, assessment tools should be utilized which most effectively measure these abilities. One practical way of achieving this objective is through surveys to assess each student's effectiveness working with classmates, team members, and the various stakeholder groups with which they collaborate during field exercises. Data from such surveys, coupled with instructor observations, could serve effectively to track student development and capacity to become an outstanding conservation professional.

There are other aspects of training conservation professionals which are of importance and should be explored. These include the type of leadership and oversight provided to the students. The typical professor tenured as a result of a lengthy publications list will be inadequate for the task at hand. It is important that instructors be dedicated to teaching first and foremost.

Universities which offer such training will need to be uniquely suited to that purpose. This would include elements such as innovative approaches to hiring faculty, establishing interdisciplinary coordination among academic departments, and the development of cooperative agreements and partnerships with diverse entities outside the university system – public, private, governmental – that would facilitate program formulation and implementation.

In wrapping up this discussion, I offer one last observation. It seems evident that most of the conservation community has yet to recognize the extent to which the training of future professionals requires overhaul. This suggests it is unlikely that a taskforce of professional conservationists charged with developing a strategy for modernizing graduate training would develop a model anything like that which is necessary. Such a discussion, therefore, would benefit from the addition of other stakeholder groups, particularly, from the most disenfranchised stakeholders I have noted in earlier chapters. Such engagement, I expect, would bridge the gap and move dialogue in a direction which, in the long-term, would benefit all parties involved – and wildlife conservation most of all.

And so we have it – a draft blueprint for a cornerstone initiative to train conservation professionals. Each stakeholder-strategy developed to reach a particular target sector of this conservation values initiative ought to have at least one. In other words, a local rancher association might create one, perhaps in collaboration with an agency such as the US Bureau of Land Management, to promote better collaboration and understanding of one-another's missions. An inner-city school district might develop a cornerstone initiative that ensures all of its third graders experience meaningful outdoor learning. A religious entity might opt to offer a comprehensive discussion series on how its sacred texts relate to treatment of the environment. And so on. Cornerstone initiatives need not address training. They might focus on other themes as these examples illustrate. The important point is that such initiatives contribute in a significant way towards engaging that stakeholder group with regard to conservation values.

CHAPTER 9

ENGAGING EVERYONE – WE ARE ALL STAKEHOLDERS

Earlier, in Chapter 6, under the subheading Broad Dialog, I mentioned the importance of engaging all of a community's members in achieving conservation. The topic was introduced earlier because of its importance. I will expand on that discussion here in exploring the final four elements of the proposed framework (elements #6, #7, #8, and #9).

FRAMEWORK ELEMENT #6:

Democratize All Conservation Practice

Most of us study the democratic process in school where we learn about the electoral process, the separation of powers, and the like. But does studying it actually mean we truly understand it? I thought I knew it reasonably well after going off to college. That was, at least, until I spent a summer with about 75 other college students from all over the Western Hemisphere from Canada to Chile and Argentina. Our task that summer was to experience democracy first-hand by creating one among ourselves. Well, that sounded easy enough, especially when the six-week program began with a nicely conceived committee system in which each student could participate to the extent that she/he desired, or bow out of the process, whichever was preferred. The committee system didn't last two weeks. Complaints came out of nowhere, to the point that the committee system was scuttled, and a meeting of all the students was held to explore other options. No problem. One of the older participants, (we were 18-24 in age) happened to be a law student, and he recited a constitution he had prepared to address our dilemma. It was an excellent piece of work. Seemed to cover all the bases. We were in business. But wait, someone had a question about one of the articles. The law student responded, but added a caveat. "I am pleased that you like this constitution," he said. "But, you should be aware that either you accept it exactly as I have written it, or you have no constitution at all." Wow. And as if that wasn't enough, he added, "And by the way, if you don't accept this constitution, I am leaving."

What? You can imagine what happened. After that meeting, we never saw the fellow again. He just disappeared. And as to a constitution, we argued nearly every night about creating one, but by the time our six weeks together was coming to a close, we were no closer to effectively governing ourselves than we had been the day the committee system was trashed. Clearly the problem wasn't just the law student. As we struggled with creating a democracy, it became increasingly obvious that other issues were involved: lust for power, deceit, misunderstandings about what a true democracy should look like, lack of trust, and translation issues (discussions were a mixture of English and Spanish).

This experience was not an isolated incident. For two subsequent summers, I worked on the staff of this incredible program, the Encampment for Citizenship, and, in each encampment, governance collapsed. Notably, in these latter two encampments, all of the participants were from the United States and so language, along with other issues, were eliminated as complicating factors.

I relate these stories to illustrate just how difficult creating a truly democratic process really is. It is amazingly challenging. And the practice of conservation, one that should engage every individual because it affects every individual, falls decidedly short in this regard on many fronts.

Broadening conservation's base requires expanding the number of individuals, communities, and societal sectors demanding that conservation be a priority. We must aim to include everyone as part of conservation's constituency. It is essential to **democratize** the process of engaging **all** stakeholders in conservation practice. This should be a fundamental principle in every conservation endeavor, but it is not. As you read through the many accounts in this book, I have touched upon various examples of non-democratic practices. I concur with Marvier and Wong (2015) that, referring to the US, but applicable to most first-world countries and international groups, "conservation organizations have systematically neglected outreach beyond their current base of white donors and members." A major reflection of this is the struggle of the hunter lobby to retain power in the face of a changing populace.

Not only is it essential that as wide a swath as possible of every community or society be engaged in this process, but such is particularly the case for entities at opposite ends of the political, social, and economic spectrums. This cannot be overemphasized. If any bias is to occur in this regard, it should be towards those who feel disenfranchised by the conservation movement. These are the private land-owners – ranchers and farmers who feel that conservationists want to dictate how they manage their land. These are residents of cities who may not see the conservation of nature as an issue of any relevance to them. These are businesses which do not recognize any connection between sustainable resource conservation and their financial bottom line. These are women who are largely marginalized by this male, hunter-dominated profession. The important challenge is for all to set aside differences and focus on commonalities of purpose. Conservation, fortuitously, has many such commonalities if we only take the time to look.

"Where there is power, there is resistance"

– MICHEL FOUCAULT

A few years ago, the state of Florida overhauled its wildlife conservation program. The state shifted from essentially focusing on animals which are hunted and fished to a much broader focus on the overall biological diversity of Florida and the underlying threats to a healthy environment, the primary threats being (1) habitat fragmentation; (2) degradation of water resources; (3) incompatible fire management; and (4) invasive plants and animals. The aim of this shift was to redirect attention towards resource issues of broader interest to that state's populace. Interestingly, this shift in focus occurred at a point when the number of registered hunters in Florida had fallen below one percent. In essence, this small group of stakeholders, up to that point, had long dominated Florida's fish and game department for themselves. Most states in the US remain where Florida had been. There are political reasons for this, mostly historical as touched upon earlier. A counterweight is needed to this imbalance. That must be a caring, concerned, engaged, and empowered public – democratizing conservation will allow the development of that public.

This situation in the states has a significant spillover into the federal government because quite a few US Fish and Wildlife Service directors are selected from state ranks. That is, perhaps, a reason why a federal conservation agency such as the US Fish and Wildlife Service has never, to my knowledge, conducted a proper survey of the US public's sentiments regarding what they would like to see as that agency's priorities. Oftentimes, it seems that the US Fish and Wildlife Service does not realize that its funding base is the federal taxpayer, unlike the states which are heavily dependent on specific taxes on hunters and anglers.

Private conservation groups are not immune to this problem.

On an international scale in particular, I, among others (Guha 1997), find neo-colonialism to be a common feature. Because it is so widely practiced and accepted, it typically goes unrecognized, even by those who might act differently if they were aware. Because it is such a sinister force, though praised on so many fronts, I chose it to begin the introduction to this book with our discussion of a billionaire purchasing land in Africa. That case is not unique; it represents a widespread practice. This issue too, has to do with our values. It is relevant to probably every important issue we face – conservation-related or otherwise. But the consequence is the same – it results in ignoring, if not subverting, core values important to society. Since effective environmental stewardship is strongly correlated with democracy (Lappe 2011), the conservation community should take special heed of this shortcoming as I believe it has a lot of soul-searching to do on this front. I cannot highlight enough that the practice of conservation should be strongly associated with a democratic process that more effectively engages its entire community. In the words of human ecologist and environmentalist Eugene Anderson (2010), "The absolutely critical point… is that this responsibility [to caring for the environment] has to add up to a strong sense of unity." In my mind, democracy and conservation are intertwined.

In the long run, the circumventing of democratic processes within the conservation community sacrifices the broad base of support necessary for it to succeed. It is also why I have **not** proposed, as so many books do, **what our values should be,** but rather suggested that it is up to each community to **determine these values for themselves.** Values must percolate from the bottom up, not be imposed from the top down.

During this discussion, we emphasized that communities at all societal levels, from the smallest civic group to federal agencies, can utilize the toolbox described here. Though the discussion focused, at some points, on the state and national levels, the intent was in no way to diminish the importance of actions at other levels. In fact, it was noted early on that a bottom-up approach, whereby small-scale initiatives build into grander ones at larger scales, would be the most democratic and powerful way for societal conservation values to grow in any country. Of fundamental importance is that there be no "agenda" involved beyond the seeking of shared conservation interests. Because there is no predetermined outcome expected of this initiative, it should be minimally threatening either to those for whom conservation appears irrelevant or those who historically have been opponents of "tree hugger" type activities.

Because building initiates from the bottom up rather than the top down is so critical, I have not proposed legislation as a mechanism to set it in motion. That is not to say that over time, legislation would not be valuable. It is just that I have come to believe that it is all too often used as a crutch, if not misused, thus it is not a critical component of how best to move forward.

A few last points concerning democracy – not only of conservation, but more generally as well. A democracy is only as strong as its weakest link. In this regard, I am referring to the capacity of a community's or a nation's populace to be engaged, knowledgeable, analytical, and generally prepared to actively participate in the democratic process. To that end, the United States erred when it discontinued the draft in 1970. Doing so deprived millions of our youth from having to leave home and be exposed to other young men, and later women, from across the country, of differing races, creeds and backgrounds, to whom they had never been exposed. Furthermore, our youth also lost the opportunity to spend time abroad in another country and be exposed, during a formative stage in their lives, to cultural difference about which they likely never even dreamed. In my mind, the loss of these experiences, experiential learning of the most important kind at a critical stage in a young person's life, created a huge setback from which this country has never recovered.

That brings me to the final point. It is not that the military draft be reinstituted. It is that the draft be substituted for by some sort of national service. Such service could take any of many forms, some of them, obviously, including initiatives to improve the environment – such as the Civilian Conservation Corps back in the 1930s. Both conservation, and democracy as a whole, would be the better for it. At this stage in US history, we have seen the pendulum swing in the direction of individual freedom. Individual freedom is not a bad thing – as long as it does not infringe upon the rights of our neighbors. But what has virtually vanished in this shift is recognition of our duty – our duty to support civil society. Democracy works best when all citizens contribute to it. Preferably, this should be voluntary – we should all want to help make our society a better place. But being voluntary likely will not work at this point, thus it should be mandatory. We sometimes need assistance in helping to do what is best for all of us.

"They who give have all things;
they who withhold have nothing." – HINDU PROVERB

FRAMEWORK ELEMENT #7:
Network by Expanding Collaboration among Communities at All Levels from Local to National

Communities do not live in a vacuum. They interact, interchange resources, and depend upon one another in various ways, whether the communities are physical neighbors, are far apart, or of completely different scales (e.g.: a town and a state).

There is great power in communities sharing with and learning from one another regarding their conservation values. Such interchanges would prove synergistic and beneficial to both parties. Over time, conferences, newsletters, websites, a registry of engaged communities, and the like focusing on this theme will greatly enhance showcasing successes and sharing of techniques to better address all elements of this framework. I foresee such interaction to be as exciting and fulfilling an activity as any conservationist could desire. To the degree that governments, the business sector, local communities, and civic organizations share together in this exchange, so much the better. That will be a sign of embracing the democratic process in a way this framework demands. Hopefully, one or more groups will step forward to facilitate communication among communities, create a website, develop a list of engaged entities, post their conservation values charters, among other actions. Enhanced networking should expand the process all the faster. Of great importance, it will also promote more consistent messaging and framing of conservation values. Obviously, the internet provides an exceptional tool in this regard, shrinking the challenges of communicating at all levels. The internet alone, with the use of social media, can serve to ignite actions discussed here at a pace undreamed of but a few decades ago.

"Coming together is a beginning, keeping together is progress, working together is success." - EDWARD E. HALE

Conservation organizations, as a community, should not only be major players in such networking, but would benefit from enhanced cooperation among themselves. Stated emphatically by Eugene Anderson (2010), "The biggest problem facing the environment today is the lack of unity among those who are most immediately dependent on, or concerned with, conservation and sane management.... Environmentalism is in sorry shape now because of too many disparate goals and too much puritanism and intolerance about getting to them." Amen. Based on extensive first-hand personal experience, I agree entirely, the conservation community is far too competitive regarding funding and recognition while short on partnership.

I'll offer a story in this regard. World Wildlife Fund – US, a major international conservation organization, had invited a number of similar entities along with myself, to a multi-day strategic planning session to, among other things, explore better collaboration among groups. To this issue, the CEO of another organization, perhaps the largest of those present, spoke. The visitor's comment? "I do not see collaboration as a way to go. I think competition is a good thing." Needless to say, he put a complete damper on the meeting, and his organization later went on to reflect his sentiments.

"Competition has been shown to be useful up to a certain point and no further, but cooperation, which is the thing we must strive for today, begins where competition leaves off." – FRANKLIN D. ROOSEVELT

On a later occasion, this same conservation group, along with several other major international conservation organizations, were invited to submit proposals to compete for a single funding grant made available by the biodiversity program in the US Department of State. Rather than compete for the funds, I hosted a meeting of the interested organizations to consider submitting only a single joint proposal to the State Department containing components particular to each of the organizations involved. This would guarantee each of them a piece of the pie to address their particular interests while reducing the effort lost in grant preparation. Development of the joint proposal went smoothly except for the interests of one group – the same one that trashed WWF's earlier efforts to promote collaboration. This group pushed relentlessly for a greater share of the pie. All the other parties ultimately relented, recognizing that the sacrifice to this group's demands was relatively small compared to the greater benefit of being ensured a portion of the funding. What happened? Unbeknownst to anyone else, this same organization, contrary to its agreement with all the other conservation groups, submitted its own separate proposal to the State Department which, it so happens, was the proposal the State Department chose to fund. Question: If a group such as this cannot honestly cooperate with its peers, how can we expect it to serve effectively to lead conservation efforts in developing countries where its hubris and authority goes unchallenged? Sadly, this group, though an extreme example, does not stand alone.

I will not elaborate on networking, or the lack of it, despite its immense importance, because it is so widely practiced in other arenas and well addressed in numerous publications. An excellent sourcebook is, Tools of Engagement: A Toolkit for Engaging People in Conservation edited by Judy Braus (2011).

FRAMEWORK ELEMENT #8:
Develop and Hire Conservation Professionals with Socially-Oriented Skill Sets

This important element has emerged frequently throughout the book. Because of its critical nature, Chapter 8 was dedicated to describing a cornerstone initiative highlighting the necessary skill-sets for modern-day professionals in the conservation field. Following on that discussion, there is no need to elaborate on this framework element. Hopefully, Chapter 8 made clear the magnitude of the transformation required in the field of the conservation professional. Applying this framework element to my own career, had I known earlier what I do today regarding the skills needed to advance conservation, I would have dramatically modified my hiring practices to select for different skill-sets in my employees. This factor alone was probably the single most important one hindering the evolution of our efforts to expand into new conservation frontiers.

Obviously the type and size of the community engaged in addressing its conservation values will influence what actions might be necessary regarding this matter. Not until a transformation takes place in our professional leadership will conservation action shift from focusing too heavily on symptoms towards addressing underlying problems. A major upheaval appears necessary on many fronts, and particularly within many states of the US before they adequately care fairly for the conservation values of **all** their citizens rather than directing immense attention to the very small minority who hunt, trap, or fish. To this point in time, such a transformation has proceeded all too slowly. For that reason, waiting for it to occur from within the profession seems futile. I believe it will need a major shake-up provoked from without. Preferably, that should come from a concerned civil society demanding action. Ultimately, developments in this profession, or lack thereof, will determine, to a large extent, what our planet looks like to generations to come.

FRAMEWORK ELEMENT #9:
Adaptively Manage and Consider Additional Conservation Values

We have now covered the fundamental elements of making values the center of conservation practice – it's a hefty toolbox. But the process is not done. Because the process is never done. In a world changing as rapidly as ours is today, and where change is only accelerating, it is unlikely there will be a time when we can simply stand pat, and presume we have reached our goal. There will always be room for improvement. Every community should seek, over time, to expand its conservation values to be increasingly compatible with the earth and all of its creatures. With experience, the challenge of developing measurable indicators and monitoring protocols for these values will be refined.

One of the most exciting potential developments would derive from communities learning from one another on this front, prompting them to go back and revise their values base to be more comprehensive. This alone would be a singularly important achievement. Also, the potential of small, innovative communities to share their successes, spread them among like-minded entities, and then, together, influence positive change at higher, more complex levels of counties, states, and nations, is particularly exciting.

Given that the conservation movement is in its relative infancy in addressing values in a meaningful way, this movement, and the communities which are its backbone, have much to learn on this front and so adaptive management of initiatives as they grow will be essential.

The process repeats itself, hopefully improving with each iteration.

CHAPTER 10

MOVING FORWARD

"I have not failed. I've just found 10,000 ways that won't work." – THOMAS EDISON

Where We Go From Here

"The most important environmental issue is one that is rarely mentioned, and that is a lack of a conservation ethic in our culture" stated Gaylord Nelson – governor, senator, and cofounder of Earth Day.

Reversing this situation is long overdue – we must heed the sages of the past. I have aimed to make the case for why their advice is so sound. Further, I offered a potential path forward. In doing so, it is important to note, I have tried not to be prescriptive. All too many authors offer a set of principles towards which society should strive. While such noble goals might be valuable, what is more important is **ownership of the process.** Whether small community or large, it is the challenge of engaging community members into the process that is more important than the immediate values derived from it. Done properly, the process builds trust. The building of trust strengthens communities. The strengthening of communities creates a foundation for loftier goals like conservation. I have come to believe firmly that the less we are invested in a specific outcome, the more open we can be to a fair, inclusive, democratic process. Only on that basis can more significant progress be made.

Is the framework I have laid out the only path? I believe all of its elements are essential. The order in which elements are implemented, however, and the degree of focus on any particular element is flexible. One option, for example, would be just to identify important conservation values that communities ought to have, but don't, and promote them in various ways such as through stakeholder-strategies or other mechanisms. This would be productive immediately. And, under some circumstances it may be the best way to start things rolling. Why not bypass all the preliminary stuff of identifying a community's existing values and developing a conservation charter? Considering all the effort involved, it may well be found that a community's conservation values are minimal and provide little to build on. Yes, that may well be true. And this short-cut approach is certainly an option.

I have proposed the more comprehensive alternative for two reasons. One, is because I believe it is especially important that communities possess a shared vision, a vision that has the power to build positive aspirations and a shared identity. Thus, I believe it is very important to clarify and memorialize conservation values. Second, is because application of the democratic process is so critical. Important to the process is not beginning with a pre-conceived idea of what values a community should have, but instead finding out and embracing the values a community already possesses. Are these the best possible values? Not likely, but they represent where the community is at the present moment. Since this framework calls for an iterative process, a community's values base can subsequently be strengthened over time. I have little doubt that utilizing the democratic process greatly enhances the potential for a community to embrace a shared vision.

Also to this point, while there may well be other effective paths for achieving conservation, there are far more that are ineffective. Well-meaning effort is lost on addressing symptoms, or using techniques demonstrated to be ineffective. As we saw with the ill-advised creation of the monarch butterfly reserve that upset local inhabitants so severely that they tried to burn it down, some seemingly beneficial conservation strategies are downright destructive.

Actually, exploring other mechanisms to incorporate societal values more effectively into conservation practice makes for a very exciting discussion. We should constantly be looking for a better roadmap to advance the incorporation of values into how conservation efforts are undertaken locally, nationally, and globally.

What I hope has become clear is that simply developing a "bigger hammer," the typical response to most dilemmas, isn't the solution to successful conservation. Conservation will not benefit from simply "more of the same." To quote another sage, Albert Einstein, "The significant problems we face cannot be solved at the same level of thinking we were at when we created them." Sticking to a long-standing formula with which we are comfortable, but which increasingly fails to solve our community's or our nation's conservation woes, is a recipe for disaster. It fits under what Gerald Zaltman, a professor of marketing at Harvard, refers to as the "Titanic Effect" (2003) – what occurs when managers have unquestioned confidence in customary assumptions about their constituents. When applied to conservation, the term has a certain irony to it considering that climate change, which includes the loss of global ice, is the greatest environmental threat of our time. Dramatically, the image captures the situation the conservation community faces: as in the case of the Titanic, we are failing to see the iceberg.

"If you only have a hammer, you tend to see every problem as a nail." – ABRAHAM MASLOW

Recently I was contacted by a colleague from a major conservation organization who has long lobbied Congress to appropriate funds for bird conservation. He requested advice on how to direct tens of millions of dollars of new money. My answer? "Don't lobby for new money. Lobby that a national survey be conducted and that existing funds be redistributed based on the conservation values identified in the survey results." That would do far more for bird conservation than pouring new money after old without some basic changes in priorities and focus of most federal bird conservation programs in the United States. This is not to say that additional funds are not needed – but both old and new funds need to be directed in more effective ways that benefit greater segments of society and our environment.

This book makes the case, increasingly supported by others, that conservation is a social phenomenon, not a biological one. That is to say, though we may see species threatened and habits destroyed, the underlying cause is human-based. We have seen that it is a society's attitudes and values, and ultimately our behaviors and norms which spring from them, that are the single most significant factor in determining whether conservation efforts succeed. Other factors matter, but not nearly to the extent to which values determine conservation success. Absent conservation values, deployment of all other factors combined – money, biological research, laws – are unlikely to succeed.

The United States and the Western World are starting at a disadvantage. As was elaborated early in this book, the ethos of the Pilgrims, and Western religions in general, have not, historically, been particularly sensitive to the intrinsic values of nature. Fine, let's accept that. What we must not accept is that because of this circumstance, we should throw our hands in the air and ignore that this problem exists. We have done that far too long. And you've seen the results. Refocusing to address our underlying problem is essential and that means centering on societal values. Does this mean that religions, eastern or western, should take over the conservation movement? No. But it does mean they should become more important and active players, especially given that many societies depend upon them for ethical and moral guidance – the basis of our values.

There will be holes in the shared values agreed upon within communities, likely gaping ones, particularly at larger scales, such as the state and national levels. But, such shortcomings serve, at the very least, to expose the challenge before us. They clear the brush and briars that impedes the long path we have ahead of us. Yes, there will be much more brush to clear, but we will finally have the satisfaction of knowing we are following the right path – a satisfaction all too absent at present.

To some extent, each value shared by a community is like a strand of thread in a fabric. Many are needed to complete the finished product. But, with the addition of each new thread, the fabric becomes stronger and stronger.

At the same time, arguably, India has demonstrated that even with only a single value – respect for all living things – the world can become a much better and more livable place for all creatures and their habitats. Fundamentally, the number of conservation related values a community possesses is far less important than the level of sacrifice that community is willing to make to ensure that its values are implemented. Sacrifice is the name of the game. Sacrifice is the ultimate measure of caring. The measure of a community's willingness to sacrifice is a much better indicator of conservation success than is the present standard of the number of acres protected. It is worth noting that as recently as 2013 India's 41 tiger reserves had over 51,000 families living in them – likely well over 200,000 people. Can anywhere in the Western World ante up this level of sacrifice in the face of life-threatening wildlife made by these Indian families?

Related to this point I need comment on the setting aside of protected areas, the continued focus of many conservation organizations. It seems the case that many people treat the setting aside of land to protect wildlife as if it were a substitute for properly treating land that is not formally protected. Sure, it is great to protect the Cape Cod National Seashore for its wildlife and scenic beauty, but does that give the Seashore's neighbor's carte blanche to spray pesticides and herbicides all over their lawns and gardens? Do they not see that the absence of purple martins and quail and marsh harriers and short-eared owls is likely the result of this mistaken attitude? It is understandable that the Seashore's neighbors may believe that biocide use is okay because they may not know any better. What is worse is that the professional managers of protected areas such as the Cape Cod National Seashore and the Monomoy National Wildlife Refuge seem to be absent in the call of duty at addressing this disastrous perception.

"A vision without a task is but a dream. A task without a vision is drudgery. A vision and a task is the hope of the world." – AN OLD CHURCH IN ESSEX, ENGLAND

Some communities, perhaps most, might lack interest in identifying their conservation values. The situation might well be similar to the one we saw with the monarch butterfly in Mexico where communities around the butterfly reserve were engaged by first addressing their more critical needs and gradually working the butterfly into the mix. Such circumstances will call for innovative approaches, but that should not prove an insurmountable obstacle. As with the butterfly, the challenge will be to begin with a conservation-related issue of immediate importance to the community. Inevitably, there are many of these. Looking at my own community on Cape Cod as an example: How do we balance the desire for more wind turbines out in Nantucket Sound with the substantial killing of migrating bird flocks by their propellers? How do we better protect our critically endangered northern right whale population, hovering around 300 animals, from ships in the channel passing through its sanctuary? How do we address the increase in great white sharks along local beaches and their threat to recreational bathers and surfers? And the list goes on. Addressing each and all of these issues ultimately revolves around the community's values and so, over time, the linkage can be made to crafting a set of overarching conservation values that can be stepped-down to guide decision-making.

While engaging societal values might seem to some, perhaps many, a distraction from coping with the tangible, everyday challenges of grappling with environmental issues, it is important to recognize that by doing so we now become proactive in addressing such concerns. As we advance on this front, we get ahead of the curve. One major concern regarding conservation as practiced today is that it is usually behind the curve and thus highly reactive. This is a poor place to be because it is a much weaker position from which to influence outcomes. To use the Puerto Rican whip-poor-will as an example, the bird I mentioned earlier that was nearly exterminated by a misplaced garbage dump – in what position would you rather be? Trying to stop the location of a dump while an agreement is about to be signed, or a permit issued, or heavy equipment unloaded to initiate digging? Or, that such a proposal be dead on arrival when first suggested because of it breaching pre-established indicators developed to ensure a community's values are not violated regarding the conservation of threatened species? I opt for the latter. The earlier that conservation concerns can be vetted in the decision-making process, the far greater the chance for success. This is no small point and one of the great challenges to environmental conservation.

We discussed the importance of conservation values in all communities, large and small, and of representing them in an informal charter, or compact of some sort. We considered how we might most effectively engage the public to embrace these values – to make them a force in the hearts of our citizens. The importance of unifying conservation messaging and best practices into a focused, consistent, well-branded, and well-marketed initiative was elaborated – an important concern both within and across communities. The use of social marketing provides a ready tool. We explored the importance of looking at society as an array of specific stakeholder groups each with its particular perspectives and sensitivities, plus we discussed the need to engage each societal sector in a comprehensive manner. We also talked about the development of a cornerstone initiative for each stakeholder group and illustrated it using the education of future conservation professionals. We addressed the essential role of integrating democratic processes in all conservation activities, the power of networking, the need for a new type of conservation professional, and the importance of adaptively managing the process as it moves forward.

Framework for Effective Community-level Conservation (civic groups, businesses, towns, state, and federal entities):

1. Identify community conservation values
2. Create a charter of community conservation values
3. Frame, implement, and monitor conservation values
4. Focus on stakeholder groups
5. Customize cornerstone initiatives for each stakeholder groups
6. Democratize all conservation practices
7. Network by expanding collaboration among communities at all levels from local to national
8. Develop and hire conservation professionals with socially-oriented skill sets
9. Adaptively manage and consider additional conservation values

Will the identification and promulgation of a community's or a nation's conservation values guarantee a dramatic shift in society's approach to conservation? Not necessarily. There is a significant step between identifying conservation values and those same values resulting in constructive actions by the community and changes in behavior by its citizens. But, I am convinced this approach offers our best chance of achieving true improvement. Of fundamental significance, it focuses on people's hearts as the most critical element of conservation, not on an individual's understanding or acceptance of scientific argument. This shifts emphasis to the true driver of human decision-making. We coupled this with providing communities a positive shared identity built around conservation concerns. There may be no more powerful a force in human society than a shared vision (Senge 1990).

Conservation values do not occur in a vacuum. Values which guide our lives, such as how we treat one another and how committed we are to a better society, are important pillars which affect the conservation approach I've laid out here. Values typically integrate and overlap, though sometimes they may conflict with each other. We might believe strongly in having a strong domestic economy, but that does not deter us from rushing off to purchase a bargain item imported from abroad, a purchase which undermines local industry. Here we have a patriotic value conflicting with another of being frugal and seeking out the least expensive product. Teasing values apart so that we better understand their place in a community may be challenging. But, the better the hierarchy of values is understood, the better the chance for achieving successful resource conservation.

As communities advance on all of the above elements, there remains the ultimate hurdle – getting the political establishment, perhaps the most powerful of all stakeholder groups, to align with their views. When all is said and done, social mobilization and political leverage carry the day. But these capacities are developed one step at a time. The processes discussed in this book should result in progress on that front as well, especially if the political stakeholder group becomes a priority focus as exemplified in Chapter 7 by the "Cabo Pulmo" decision-maker case in Mexico.

"The greater the power, the more dangerous the abuse." - EDMUND BURKE

We briefly discussed the pace at which attitudes and values can change. It seems evident that this process is accelerating daily, a transformation which we have no choice but to accept. We must not only accept it, but we must embrace it and use it to conservation's advantage.

"Change is inevitable. Progress is optional."

– TONY ROBBINS

An immediate challenge is to strengthen recognition within the conservation community of the importance of values as a conservation tool. In fact, the conservation community, in actuality, needs a culture change in this regard. That is why such an extensive component of the book was focused specifically on demonstrating its importance. Not only must the conservation community's focus be enhanced, but it must seek alignment across the movement to strengthen synergy among groups and improve framing of issues and messaging to the public. Presently, resource conservation is practiced in such a disparate manner, each group doing its own thing, that it scarcely seems like a movement at all. What are its shared values that all groups identify with and promote? What is its golden rule? What unifies it as a movement? What defines it? I believe that as the conservation community moves forward on these fronts, it will take a major step forward. The conservation community must consolidate itself as a movement while engaging as a central participant in all the actions identified for other communities (civic groups, towns, etc.).

We discussed that addressing conservation in this way is not what most conservationists and resource managers are trained to do, nor are comfortable doing. As a result, many will declare this approach impractical, of secondary importance or, perhaps of long-term significance, but not timely, given the urgency of the problem and the limited resources at hand. Until a transformation takes place, conservation professionals will be their own worst enemy.

Regrettably, expansive and highly hyped U.S. based graduate-level training in wildlife conservation and other environmental fields has led numerous students from other countries to come to the U.S. to learn "state of the art" skills in this field. Conversely, how many Americans seek such training in India? Probably zero. Conversations with Indian associates suggest that anywhere from 30-50 percent of that nation's Ph.D. professionals are trained in the United States. I would argue that global conservation would be much better served if that trend were reversed and graduate students sought out India as the mecca for learning how to achieve wildlife conservation. I say this with the caveat that India must first recognize itself the importance of local values and stop adopting flawed approaches from abroad.

Some may consider this proposal too radical. Well, it may seem that way. But it only appears that way because we have ignored for too long the need to focus on broader society's conservation values. The consequence is that we have let other interests mold our culture into one increasingly removed from nature and its importance to us. Such is the case not only for the US, but for nearly every nation on earth. Despite this unfortunate societal shift away from respecting the primary source of all the goods and pleasures upon which we now depend, I believe the ground can be made up rather quickly to get us back on track. We have a greater potential to succeed than we might think. We have every reason to be hopeful because we have a number of powerful resources at our disposal.

"Nobody can go back and start a new beginning, but anyone can start today and make a new ending."
– MARIA ROBINSON

We have youth on our side, as the young are the vanguard of social change. A focus on conservation values would provide them a noble, substantive cause towards which to dedicate their enormous vigor and idealism. There could be no better champions. Also, since our childhood experiences are the single most important factor in determining our values and most of the behaviors we will adopt throughout our lives, the engagement of youth in every way possible is central to this endeavor. After all, our greatest connection to the future is through youth.

Upon graduation from college, I joined the National Teacher Corps. This federal program, now discontinued, was something like a national Peace Corp, but focused on educating disadvantaged youth. During my service in Gary, Indiana, I likely learned more than did the students with whom I interacted. But much of it was not positive. Far too many of my tenth-graders had virtually never been to a natural area. Thus, many believed that a harmless spotted salamander we found on one of our field trips was deadly poisonous. Several believed that women could have children without intercourse. Many had no idea of the celestial basis for a month or a year on our calendar. Little has occurred since my time in the National Teacher Corps to suggest the situation has changed. In fact, the renowned commentator Bruce Jennings, many decades later, in a special feature on American values, highlighted the school in which I taught, and all indications were that things had deteriorated since my time there. Regardless, my point is not to denigrate inner-city schools, far from it. It is to suggest that for all intents and purposes these students had been ignored by society.

As such, they are disconnected from many aspects of learning and civic engagement – and that includes recognizing and being involved in supporting conservation. Neither the conservation community specifically, nor the nation as a whole, gains from this. Fortuitously, there have been many advances on this front in recent years. A stunning example is the No Child Left Indoors initiative, a coalition of over 2,000 organizations dedicated to addressing this very problem. Despite these advances, there is much more to be done, consequently, youth in every community must be a major focus of moving this conservation initiative forward.

Especially exciting is the capacity to offer citizens of any community, large or small, the opportunity to develop a noble goal or mission – to give a powerful, positive meaning to people's lives. In conquering our basic needs, modern society has stripped many of us of having purposeful goals towards which to dedicate ourselves. We all need such goals. They are sorely lacking today. The identification and fostering of conservation values has the potential to fill this void and ennoble all of us. Research has found that once people begin to see themselves as concerned citizens, they tend to act increasingly like one.

A suite of conservation values, as promulgated here, ultimately should be an important part of each person's philosophy of living, their worldview. To some, such values may even prove spiritual or religious. Looking at conservation from such perspectives as these is a positive development and reflects how differently we have to focus our efforts. We can do much worse than have a religion focused on how humans relate to planet earth. Arguably, the broad scale adoption of a modern religion with this focus is long overdue.

I believe identifying, memorializing, and integrating societal conservation values into community dialogue and action is practical and achievable because the issue is so one-sided. Who does not want an earth that will provide for our children and grandchildren? Who would deliberately choose to leave behind a scorched earth attractive to no one? The challenges of implementing this approach do not diminish the need for us to undertake the task. It is the only choice available to us that seems to have real potential to reverse the dire situation in which we find ourselves with regard to the natural world.

This approach is applicable globally. People in all nations suffer similar environmental threats and need effective paths towards conservation. I have written this narrative in the context of the United States because the US sets trends that many nations follow. That said, there is no reason why initiatives to promote societal conservation values could not be developed anywhere in the world. In actuality, given our earlier discussion, India would be a most exciting country to see such efforts transpire. After all, India has led all nations in wildlife conservation. In that regard, the world owes India a great debt of gratitude for the sacrifices its people have made and continue to make to sustain populations of some of the world's most spectacular, but dangerous animals.

Various international treaties and conventions, particularly environmental ones, should implement this process as practical with regard to their mandates. These bodies can also help promote conservation values-related activities in their member countries.

The values-oriented approach proposed here with a focus on wildlife conservation is, I believe, applicable to other environmental fields. The climate change debate, discussed earlier, is a prime example of this. With reasonable likelihood, all other issues, conflictive or otherwise, which revolve around human actions and behaviors could benefit from review under the lens of societal values.

Returning to our imagining of the revoyage of the *Mayflower* as a ship that transported emigrants from India rather than Pilgrims to the shores of North America, we saw how differently the environs of Plymouth Rock might look today were the settlers of a different worldview. This parable vividly reflects Lynn White's (1967) profound comment, "More science and more technology are not going to get us out of the present ecologic crisis until we find a new religion, or rethink our old one." Indeed, we need to rethink our values. Upon their arrival in the New World, the Pilgrims drafted the *Mayflower Compact*. Now, 404 years later, it is time we create new conservation compacts. Compacts developed individual by individual, community by community, and nation by nation.

Earth Day, celebrated in April, first coalesced half a century ago around environmental values, poorly recognized at the time, but brought to the surface in an event of momentous consequence. This celebration continues to this day as a beacon reflecting the better angels of humanity. We can and we must do better than this. It is long overdue that we celebrate our Earth not once a year, but on a daily basis. **Every day** must be Earth Day – a celebration of our conservation values. It is only in this way that the dramatic changes necessary to restore our tormented planet can be achieved. Hopefully, the chapters you have read here provide the groundwork for achieving just that.

Our world is permeated by values. They are changing every day – faster and faster – some for the better, others for the worse. Arguably, we are in a stealth war over values. We must embrace the opportunity to influence this process constructively. It is time we coalesce our shared conservation values and use the bonding power of these unifying ideals to create a brighter future of which we and future generations will be proud. ■

LITERATURE CITED

Agarwal, A. "Can Hindu Beliefs and Values Help India Meet Its Ecological Crisis?" In *Hinduism and Ecology: The Intersection of Earth, Sky, and Water,* eds. C. K. Chapple and M. E. Tucker, 165-79. Cambridge: Harvard University Press, 2000.

Aggarwal, S. Kolam: *The Soul of Tamil Cultural and Social Life.* U.K.: Lonely Planet, 2000..

Akerlof, K. and C. Kennedy. *Nudging Toward a Healthy Natural Environment: How Behavioral Change Research Can Inform Conservation.* Virginia: George Mason University, 2013.

Alley, K. D. "Idioms of Degeneracy: Assessing Ganga's Purity and Pollution." In *Worldviews, Religion, and the Environment: A Global Anthology,* ed. R. C. Foltz. 143-160. Belmont, CA: Wadsworth, Cengage Learning, 2008.

Anderson, E. N. *Ecologies of the Heart.* New York: Oxford University Press, 1996.

Aubel, J. "Elders: A Cultural Resource for Promoting Sustainable Development." In *State of the World: Transforming Cultures,* The Worldwatch Institute, 41-46. New York: W. W. Norton, 2010.

Barry, D. and M. Oelschlaeger. "A Science for Survival: Values and Conservation Biology." *Conservation Biology* (1996) 10:905-911.

Bekoff, M. "Who Lives, Who Dies, and Why: It Shouldn't Be All About Us." In *Ignoring Nature No More: The Case for Compassionate Conservation,* ed. M. Bekoff, xiii-xxvii. Chicago: University of Chicago Press, 2013

Bekoff, M. *Rewilding Our Hearts: Building Pathways of Compassion and Coexistence.* California: New World Library, 2014.

Berry, W. "A Promise Made in Love, Awe and Fear." In *Moral Ground: Ethical Action for a Planet in Peril,* eds. K. D. Moore and M. P. Nelson, 387-9. San Antonio: Trinity University Press, 2010.

Berwick, S. H. and V. B. Saharia. *The Development of International Principles and Practices of Wildlife Research and Management: Asian and American Approaches.* Delhi: Oxford University Press, 1995.

Bhasam, A. L. *The Wonder that Was India.* Calcutta: Rupa and Co., 1954.

Bowen, Howard. The Social Responsibilities of the Businessman. University of Iowa Press, 1953.

Braus, J. (ed.) *Tools of Engagement: A Toolkit for Engaging People in Conservation.* National Audubon Society, 2011.

Burns, J. F., "In India, Attacks by Wolves Spark Old Fears and Hatreds." *New York Times,* September 1, 1996.

Callicott, J. B. *Earth's Insights: A Multicultural Survey of Ecological Ethics from the Mediterranean Basin to the Australian Outback.* Berkeley: University of California Press, 1994.

Chaleff, I. *The Courageous Follower: Standing Up to and For Our Leaders.* San Francisco: Barrett-Koehler Publishers, 2009.

Child, M. "The Thoreau Ideal as a Unifying Thread in the Conservation Movement." Conservation Biology (2009) 23(2):241-43.

Choudhury, A. "Human-Elephant Conflicts in Northeast India." Human Dimensions of Wildlife (2004) 9:261-70.

Clayton, S. and G. Myers. *Conservation Psychology: Understanding and Promoting Human Care for Nature.* UK: Wiley-Blackwell, 2009.

Cooney, N. *Change of Heart: What Psychology Can Teach Us about Spreading Social Change.* New York: Lantern Books, 2011.

Costello, C., S. Gaines, and L. R. Gerber. "A Market Approach to Saving the Whales." Nature (2012) 481:139-40.

Corbett, J. *The Man-eating Leopard of Rudraprayag.* Oxford: Oxford University Press, 1947.

Dee, J. "The myth of '18 to 34.'" *SPAN XLV* (2004) 3:36-39.

Divyabhanusinh, *The End of the Trail: The Cheetah in India.* New Delhi: Banyan Books, 1995.

Doak, D. F., V. J. Bakker, B. E. Goldstein, and B. Hale. "What Is the Future of Conservation?" In *Protecting the Wild: Parks and Wilderness, the Foundation for Conservation,* eds. G. Wuerthner, E. Crist, and T. Butler, 27-35. Washington, DC: Island Press, 2015.

Dwivedi, O. P. "Dharmic Ecology." In *Hinduism and Ecology: The Intersection of Earth, Sky, and Water,* eds. C. K. Chapple and M. E. Tucker, 3-22. Cambridge: Harvard University Press, 2000.

Dwivedi, O. P. and B. Tiwari. *Environmental Crisis and Hindu Religion.* New Delhi: Gitanjali Publishing House, 1987.

Elder, J. "Murky Waters: When There's No Clear Line Between the Right and Wrong Choices. In *Ethics for a Small Planet: A Communications Handbook on the Ethical and Theological Reasons for Protecting Biodiversity.* Biodiversity Project, 68-9. Madison: Biodiversity Project, 2002.

Ehrenfeld, D. "The Making of Conservation Biology." *Conservation Biology* (1993) 7(4):743-45.

Evans, D. M., J. P. Che-Castaldo, D. Crouse, F. W. Davis, R. Epanchin-Niell, C. H. Flather, R. K. Frohlich, D. D. Goble, Y. Li, T. D. Male, L. L. Master, M. P. Moskwik, M. C. Neel, B. R. Noon, C. Parmesan, M. W. Schwartz, J. M. Scott, and B. K. Williams. "Species Recovery in the United States: Increasing the Effectiveness of the Endangered Species Act." *Issues in Ecology* (2016) No. 20.

Fears, D. "Federal Judge Blasts Fish and Wildlife Service, Says Endangered Wolves Cannot Be Shot." *Washington Post,* November 5, 2018.

Flavin, C. "Preface." In *State of the World: Transforming Cultures,* The Worldwatch Institute, xvii-xix. New York: W. W. Norton, 2010.

Foltz, R. C. "Understanding Our Place in a Global Age." In

Worldviews, Religion, and the Environment: A Global Anthology, ed. R. C. Foltz, 1-7. Belmont, CA: Wadsworth, Cengage Learning, 2008.

Fox, C. H. "Coyotes, Compassionate Conservation, and Coexistence." In *Ignoring Nature No More: The Case for Compassionate Conservation*, ed. M. Bekoff, 119-24. Chicago: University of Chicago Press, 2013.

Gardner, G. "Engaging Religions to Shape Worldviews." In *State of the World: Transforming Cultures*, The Worldwatch Institute, 23-29. New York: W. W. Norton, 2010.

Gardner, G. T. and P. C. Stern. *Environmental Problems and Human Behavior.* Boston: Pearson Custom Publishing, 2002.

Gerstner, L. V. *Who Says Elephants Can't Dance?* New York: Harper Business, 2002.

Geist, V. "Noah's Ark II: Rescuing Species and Ecosystems." In *Ethics on the Ark: Zoos, Animal Welfare, and Wildlife Conservation*, ed. B. G. Norton, M. Hutchins, E. F. Stevens, and T. L. Maple, 93-101. Smithsonian Institution Press, 1995.

Geist, V., S. P. Mahoney, and J. F. Organ. "Why Hunting Has Defined the North American Model of Wildlife Conservation." *Transcripts of the North American Wildlife and Natural Resources Conference* (2001) 66:175-85.

Geist, V. "The North American Model of Wildlife Conservation: A Means of Creating Wealth and Protecting Public Health While Generating Biodiversity." In *Gaining Ground: In Pursuit of Ecological Sustainability*, ed. D.M. Lavigne, 285-93. International Fund for Animal Welfare, Canada: University of Limerick, 2006.

Gigliotti, L. M. D. L. Shroufe, and S. Gurtin. "The Changing Culture of Wildlife Management." In *Wildlife and Society: The Science of Human Dimensions.* eds. M.J. Manfredo, J.J. Vaske, P. J. Brown, D. J. Decker, and E. A. Duke, 75-89. Washington,

DC: Island Press, 2009.
Gladwell, M. *The Tipping Point.* New York: Little, Brown and Co., 2002.
Gladwell, M. *Outliers: The Story of Success.* New York: Back Bay Books, 2008.
Goodall, J. "Caring for People and Valuing Forests in Africa." In *Protecting the Wild: Parks and Wilderness, the Foundation for Conservation,* eds. Wuerthner, G., E. Crist, and T. Butler, 21-26. Washington, DC: Island Press, 2015.
Guha, R. "The Authoritarian Biologist and the Arrogance of Anti-humanism: Wildlife Conservation in the Third World. *The Ecologist* (1997) 27:14-20.
Guttmann, A. US Advertising Industry – Statistics and Facts. *Statista,* April 7, 2020.
Hance, J. "Over 2,500 Wolves Killed in US's Lower 48 Since 2011." *Mongabay News,* January 28, 2014.
Hawken, P. "The Most Amazing Challenge." In *Moral Ground: Ethical Action for a Planet in Peril,* eds. K. D. Moore and M. P. Nelson, 463-68. San Antonio: Trinity University Press, 2010.
Heberlein, T. A. "Navigating Environmental Attitudes." *Conservation Biology* (2012) 26:583-85.
Heberlein, T. A. *Navigating Environmental Attitudes.* Oxford: Oxford University Press, 2012.
Heath, C. and D. Heath. *Switch: How to Change Things When Change Is Hard.* New York: Broadway Books, 2010.
Herrera, A. O., H. D. Scolnik, G. Chichilnisky, G. C. Gallopin, J. E. Hardoy, D. Mosovich, E. Oteiza, G. L. de Romero Brest, C. E. Suarez, and L. Talavera. *Catastrophe or New Society? A Latin American World Model.* Ottawa: International Development Research Centre, 1976.
Hickman, T. and J. Ward. "The Dark Side of Brand Community: Inter-group Stereotyping, Trash Talk, and Schadenfreude."

Advances in Consumer Research (2007) 34:314-19.
Hinton, J. W., G. C. White, D. R. Rabon, and M. J. Chamberlain. "Survival and Population Size Estimates of the Red Wolf." *Journal of Wildlife Management* (2017) 81(3):417.
Holland, J. S. "The Plight of the Honeybee." *National Geographic.* May 13, 2013. https://www.nationalgeographic.com
Howe, J. *Crowdsourcing: Why the Power of the Crowd Is Driving the Future of Business.* New York: Three Rivers Press, 2009.
Infeld, M. and A. Mugisha. *Integrating Cultural, Spiritual and Ethical Dimensions into Conservation Practice in a Rapidly Changing World.* MacArthur Foundation position paper, 2000.
Jain, P. "Bishnoi: An Eco-theological 'New Religious Movement' in the Indian Desert." *Journal of Vaishnava Studies* (2010) 19(1):1-20.
Kellert, S. R. "Social and Perceptual Factors in the Preservation of Animal Species. In *The Preservation of Species,* ed. B. G. Norton, 50-73. Princeton: Princeton University Press, 1986.
Kellert, S. R. "For the Love and Beauty of Nature." In *Moral Ground: Ethical Action for a Planet in Peril,* eds. K. D. Moore and M. P. Nelson, 373-78. San Antonio: Trinity University Press, 2010.
Kennedy, J. J. and J. W. Thomas. "Managing Natural Resources as Social Value." In *A New Century for Natural Resources Management,* eds. Knight, R. L. and S. F. Bates. Washington, DC: Island Press, 1995.
Kotter, J. P. and D. S. Cohen. *The Heart of Change.* Boston: Harvard Business School Press, 2002.
Lakoff, G. *Don't Think of an Elephant!* White River Junction: Chelsea Green Publishing, 2004.
Lappe, F. L. *EcoMind: Changing the Way We Think, to Create the World We Want.* New York: Nation Books, 2011.
Lee, D. "The Natural History of the Ramayana." In *Hinduism*

and Ecology: The Intersection of Earth, Sky, and Water, eds. C. K. Chapple and M. E. Tucker, 245-68. Cambridge: Harvard University Press, 2000.

Lee-Ashley, M., the CAP Public Lands Team, and the CAP Oceans Team. *How Much Nature Should America Keep?* Center for American Progress, Energy and Environment. 2019.

Leopold, A. "The State of the Profession." *Journal of Wildlife Management* (1940) 4:343-46.

Leopold, A. "Wildlife in American Culture." *Journal of Wildlife Management* (1943) 7(1):1-6.

Leopold, A. *A Sand County Almanac and Sketches Here and There.* New York: Ballantine Books, 1949.

Levin, A. M. "Contrast and Assimilation Processes in Consumers' Evaluation of Dual Brands." *Journal of Business and Psychology* (2002) 17:145-54.

Li, P. J. "Explaining China's Wildlife Crisis: Cultural Tradition or Politics of Development." In *Ignoring Nature No More: The Case for Compassionate Conservation,* ed. M. Bekoff, 317-30. Chicago: University of Chicago Press, 2013.

Linn, S. "Commercialism in Children's Lives." In *State of the World: Transforming Cultures.* New York: W. W. Norton, 2010.

Louv, R. *Last Child in the Woods: Saving Our Children from Nature-deficit Disorder.* Chapel Hill: Algonquin Books, 2009.

Manfredo, M. J. *Who Cares About Wildlife? Social Science Concepts for Exploring Human-wildlife Relationships and Conservation Issues.* New York: Springer Science + Business Media, 2008.

Manfredo, M. J., T. L. Teel, and H. Zinn. "Understanding Global Values Toward Wildlife." In *Wildlife and Society: The Science of Human Dimensions,* eds. M. J. Manfredo, J. J. Vaske, P. J. Brown, D. J. Decker, and E. A. Duke, 31-43. Washington, DC: Island Press, 2009.

Marvier, M. and H. Wong. "Move Over, Grizzly Adams:

Conservation for the Rest of Us." In *After Preservation: Saving American Nature in the Age of Humans,* eds. B. A. Minteer and S. J. Pyne, 170-177. Chicago: University of Chicago Press, 2015.

Max, D. T. "Green Is Good." *The New Yorker,* May 12, 2014.

McKenzie-Mohr, D. *Fostering Sustainable Behavior.* Canada: New Society Publishers, 2011.

McKibben, B. "Something Braver than Trying to Save the World." In *Moral Ground: Ethical Action for a Planet in Peril.* eds. K. D. Moore and M. P. Nelson, 174-77. San Antonio: Trinity University Press, 2010.

Menon, V. "A Triangular Playing Field: The Social, Economic, and Ethical Context of Conserving India's Natural Heritage." In *Ignoring Nature No More: The Case for Compassionate Conservation,* ed. M. Bekoff, 331-42. Chicago: University of Chicago Press, 2013.

Miaoulis, N. J. "Right v. Right Conflicts: A Process for Ethical Decision Making." In *Ethics for a Small Planet: A Communications Handbook on the Ethical and Theological Reasons for Protecting Biodiversity.* Biodiversity Project, 70-73. Madison, WI: Biodiversity Project, 2002

Minteer, B. A. and S. J. Pyne. "Writing on Stone, Writing in the Wind." In *After Preservation: Saving American Nature in the Age of Humans,* eds. B. A. Minteer and S. J. Pyne, 1-8. Chicago: University of Chicago Press, 2015.

Moore, K. D. and M. P. Nelson. "Toward a Global Consensus for Ethical Action." In *Moral Ground: Ethical Action for a Planet in Peril.* eds. K. D. Moore and M. P. Nelson, xv-xxiv. San Antonio: Trinity University Press, 2010.

Nagarajan, V. "Rituals of Embedded Ecologies: Drawing Kolams, Marrying Trees, and Generating Auspiciousness." In *Hinduism and Ecology: The Intersection of Earth, Sky, and Water,* eds. C. K. Chapple and M. E. Tucker, 453-68. Cambridge: Harvard

University Press, 2000.

Narayanan, V. "Water, Wood, and Wisdom: Ecological Perspectives from the Hindu Traditions." In *Worldviews, Religion, and the Environment: A Global Anthology,* ed. R. C. Foltz, 130-43. Belmont, CA: Wadsworth, Cengage Learning, 2008.

Nelson, M. P. "The Ways We Value Nature." In *Ethics for a Small Planet: A Communications Handbook on the Ethical and Theological Reasons for Protecting Biodiversity.* Biodiversity Project, 56-60. Madison, WI: Biodiversity Project, 2002.

Nelson, M. P., J. A. Vucetich, P. C. Paquet, and J. K. Bump "An Inadequate Construct: North American Model: What's Flawed, What's Missing, What's Needed." *The Wildlife Professional* (2011) Summer: 58-60.

Nickerson, R. S. *Psychology and Environmental Change.* London: LEA Publishers, 2003.

Norton, B. G. *Why Preserve Natural Variety?* Princeton: Princeton University Press, 1987.

Norton, B. G. *Sustainability: A Philosophy of Adaptive Ecosystem Management.* Chicago: University of Chicago Press, 2005.

Oakes, J. *The Ruling Race: A History of American Slaveholders.* New York: W. W. Norton, 1998.

Olson, R. *Don't Be Such a Scientist: Talking Substance in an Age of Style.* Washington, DC: Island Press, 2009.

Orr, D. W. "Is Conservation Education an Oxymoron?" *Conservation Biology:* (1990) 4(2): 119-21.

Orr, D. W. "For the Love of Life." *Conservation Biology:* (1992) 6(4): 486-87.

Orr, D. W. *The Nature of Design: Ecology, Culture, and Human Intention.* New York: Oxford University Press, 2002.

Orr, D. W. "What is Higher Education for Now?" In *State of the World: Transforming Cultures,* The Worldwatch Institute, 75-82.

New York: W. W. Norton, 2010.
Patterson, K., J. Grenny, D. Maxfield, R. McMillan, and A. Switzler. *Influencer: The Power to Change Anything.* New York: McGraw-Hill, 2008.
Penland, M. "Our Leaders' Core Values, Revealed by Sequestration." *Washington Post,* July 17, 2013.
Perlman, D. L. and G. Adelson. *Biodiversity: Exploring Values and Priorities in Conservation.* Cambridge: Harvard University Press, 1997.
Pramling Samuelsson, I. and Y. Kaga. "Early Childhood Education to Transform Cultures for Sustainability." In *State of the World: Transforming Cultures,* The Worldwatch Institute, 57-61. New York: W. W. Norton, 2010.
Rajpurohit, K. S. "Child Lifting: Wolves in Hazaribagh, India." *Ambio* (1999) 28(2):162-66.
Rangarajan, M., A. Desai, R. Sukumar, P.S. Easa, V, Menon, S. Vincent, S. Ganguly, B.K. Talukdar, B. Singh, D. Mudappa, S. Chowdhary and A.N. Prasad. *Gajah. Securing the Future for Elephants in India. The Report of the Elephant Task Force, Ministry of Environment and Forests.* New Delhi, India: Ministry of Environment and Forests, August 31, 2010. http://www.environmentandsociety.org/node/2697.
Rawles, K. "A Copernican Revolution in Ethics." In *Moral Ground: Ethical Action for a Planet in Peril,* eds. K. D. Moore and M. P. Nelson, 88-95. San Antonio: Trinity University Press, 2010.
Rolston, H. *Conserving Natural Value.* New York: Columbia University Press, 1994.
Rosenberg, K. V., A. M. Dokter, P. J. Blancher, J. R. Sauer, A. C. Smith, P. A. Smith, J. C. Stanton, A. Panjabi , L. Helft, M. Parr, and P. P. Marra. "Decline of the North American Avifauna." *Science* (2019) 10.1126/science.aaw1313.

Rukmani, T. S. "Literary Foundations for an Ecological Aesthetic: Dharma, Ayurveda, the Arts, and Abhijnanasakuntalam." In *Hinduism and Ecology: The Intersection of Earth, Sky, and Water,* eds. C. K. Chapple and M. E. Tucker, 101-25. Cambridge: Harvard University Press, 2000.

Sachs, J. and S. Finkelpearl. "From Selling Soap to Selling Sustainability: Social Marketing." In *State of the World: Transforming Cultures,* The Worldwatch Institute, 151-156. New York: W. W. Norton, 2010.

Sandel, M. J. *Democracy's Discontent: America in Search of a Public Philosophy.* Belknap Press, Cambridge, 1996.

Sarma, L. "Significance of Kolams." *eSamskriti.* January 20, 2019. https://www.esamskriti.com/e/Culture/Indian-Culture/Significance-of-KOLAMS--1.aspx

Saunders, C. D., A. T. Brook, and O. E. Myer. "Using Psychology to Save Biodiversity and Human Well-being." *Conservation Biology* (2006) 20(3): 702-05.

Saylan, C. and D. T. Blumstein, *The Failure of Environmental Education (and How We Can Fix It).* Berkeley: University of California Press, 2011.

Schaller, G.B. "Behind the Scenes." *Wildlife Conservation* (1995) 98(3):2.

Schaller, G. B. "Politics Is Killing the Big Cats." *National Geographic:* December 2011: 89-91.

Schor, J. B. *Born to Buy: The Commercialized Child and the New Consumer Culture.* Simon and Schuster, 2005.

Schultz, P. W. and L. Zelezny. "Reframing Environmental Messages to Be Congruent with American Values." *Human Ecology.* (2003) 10(2): 126-36.

Senge, P. M. *The Fifth Discipline.* New York: Currency Doubleday, 1994.

Seshagiri Rao, K. L. "The Five Great Elements

(Pancamahabhuta): An Ecological Perspective." In *Hinduism and Ecology: The Intersection of Earth, Sky, and Water,* eds. C. K. Chapple and M. E. Tucker, 23-38. Cambridge: Harvard University Press, 2000.

Sewell, B. "Bringing Back the Fish: An Evaluation of US Fisheries Rebuilding under the Magnuson-Stevens Fishery Conservation and Management Act." *Natural Resources Defense Council Report,* 2013.

Sharma, K. "Corporate Social Responsibility- An Undeniable Helper amidst the Covid-19 Pandemic." *International Journal of Law Management & Humanities* (2020) 3(4):29.

Sharma, D. K., J.E. Maldonado, Y.V. Jhala,, and R.C. Fleischer "Ancient Wolf Lineages in India." *Proceedings of the Royal Society B: Biological Sciences.* (2004) 271(Suppl 3): S1–S4.

Singh, H. S. and L. Gibson. "A Conservation Success Story in the Otherwise Dire Megafauna Extinction Crisis: The Asiatic Lion *(Panthera leo persica)* of Gir Forest." *Biological Conservation* (2011) 44(5):1753-57.

Social Capital Project of The Resource Innovation Group, *The Ecological Roadmap: A Guide to American Social Values and Environmental Engagement.* Earthjustice, December 2008.

Soule, M. E., and B. A. Wilcox. *Conservation Biology: An Evolutionary-ecological Perspective.* Sunderland, MA: Sinauer Associates, 1980.

Speth, J. G. "The Limits of Growth." In *Moral Ground: Ethical Action for a Planet in Peril,* eds. K. D. Moore and M. P. Nelson, 3-8. San Antonio: Trinity University Press, 2010.

Steinberg, K. *The Harris Poll #41,* Harris Interactive, April 18, 2012.

Thomas, J. W. *The Forest Service Ethics and a Course to the Future.* Washington, DC: United States Department of Agriculture Forest Service, 1995.

Tinker, G. "An American Indian Cultural Universe." In *Moral Ground: Ethical Action for a Planet in Peril,* eds. K. D. Moore and M. P. Nelson, 196-201. San Antonio: Trinity University Press, 2010.

Tucker, M. E. "Ecological Themes in Taoism and Confucianism." In *Worldviews, Religion, and the Environment: A global Anthology,* ed. R. C. Foltz, 217-23. Belmont, CA: Wadsworth, Cengage Learning, 2008.

Varghese, A. and Y.T. Uhls. *The Rise and Fall of Fame: Tracking the Landscape of Values Portrayed on Tween Television from 1997-2017.* Los Angeles: UCLA Center for Scholars and Storytellers.

Vucetich, J.A. and M.P. Nelson. "The Infirm Ethical Foundations of Conservation." In *Ignoring Nature No More: The Case for Compassionate Conservation,* ed. M. Bekoff, 9-26. Chicago: University of Chicago Press, 2013.

Walker, J. F. *Ivory Ghosts.* New York: Grove Press, 2009.

Wang, L., Y-P. Ma, Q-J. Zhou, Y-P. Zhang, P. Savolainen, and G-D. Wang. "The Geographical Distribution of Grey Wolves (*Canis lupus*) in China: A Systematic Review." *Zoological Research* (2016) 37(6) 315-326.

Weir, J. "The Sweetwater Rattlesnake Round-up: A Case Study in Environmental Ethics." *Conservation Biology* (1992) 6(1): 116-127.

White Jr., L. "The Historical Roots of Our Ecological Crisis." *Science* (1967) 155:1203-07.

Winter, D. D., and S. M. Koger. *The Psychology of Environmental Problems.* New York: Psychology Press, 2004.

Zaltman, G. *How Customers Think: Essential Insights into the Mind of the Market.* Cambridge: Harvard Business School Press, 2003.

WEBSITES

Avibird. https://avibirds.com/pink-headed-duck/

BBC News. https://www.bbc.com/news/world-asia-india-47317361

Charity Navigator. https://www.charitynavigator.org/index.cfm?bay=content.view&cpid=42

Defenders of Wildlife. https://defenders.org

American Forester. https://www.eforesters.org

Facts and Details: India. www.factsanddetails.com/india/

Gallup Poll. 2015. http://www.gallup.com/poll/1615/Environment.aspx

The Hindu. https://www.thehindu.com/sci-tech/energy-and-environment/in-2019-95-tiger-deaths-in-india

Indian Country Today. https://indiancountrytoday.com

Interagency Grizzly Bear Committee. www.igbconline.org

International Crane Foundation. https://www.savingcranes.org

International Union for Conservation of Nature. https://www.iucn.org/commissions/ssc-groups/mammals/mammals-a-e/cat

International Wolf Center. https://wolf.org

Mountain Lion Foundation. https://www.mountainlion.org

National Wildlife Federation. https://www.nwf.org

National Wildlife Health Center. https://www.usgs.gov/centers/nwhc

Protected Area Downgrading, Downsizing, and Degazetting. https://www.conservation.org/projects/paddd/

Red Wolf Coalition. https://redwolves.com

Smithsonian Institution – National Zoo. https://nationalzoo.si.edu/animals/american-bison

United States Fish and Wildlife Service. https://fws.gov

The Victorian Military Society. https://www.victorianmilitary.org

Walk Through India. www.walkthroughindia.com

Wildlife Conservation Society. https://www.wcs.org

Wildlife Institute of India. https://wii.gov.in

Wolf Worlds. https://www.wolfworlds.com

World Population Review. https://worldpopulationreview.com

ACKNOWLEDGMENTS

This book would never have been completed were it not for the strong and continual encouragement of Krishna Roy, Kathleen Rogers, and Steven Stone. I cannot thank them enough. Others to whom I owe gratitude include Susan Bass, Paul Butler, Gilberto Cintron, Jonathan Cobb, Barbara Dean, Rhonda Forristall, Ruben Freyre, Katie Frohardt, Peter Howard, Larry Mason, Barbara Mathews, Teiko Saito, and Jack Turnell. For the opportunity to have an unimaginably fulfilling career I thank the Department of Natural Resources of Puerto Rico and the US Fish and Wildlife Service. The cover image and design were created by the imaginative efforts of Mariya Prytula who also contributed substantial constructive advice regarding the manuscript and book production. Kate Victory Hannisian and Claire Herndon provided valuable edits to the manuscript. Mindy Benham designed the book interior and was of great assistance with procuring photos and other aspects of the book. I especially thank my wife, Jan, for putting up with this long, arduous process and offering both the most detailed and profound edits to the final drafts.

Photo Credits

American bison – US Fish and Wildlife Service Photo Library (USFWS); Asian elephant – Popovatetiana, Adobe Stock; black-footed ferret – J. Michael Lockhart, USFWS; Cheetah – Pixabay; Deep River elephant statue – Deep River Historical Society; Deep River ivory products – Deep River Historical Society; Eskimo curlew – USFWS; gray wolf – USFWS; great auk – Kate Hollamby; grizzly bear – Casey Anderon, Flickr; heath hen – Biodiversity Heritage Library; kolam – Pixabay; leopard – ArtushFoto, Adobe Stock; lion – Ralph Lear, Adobe Stock; monarch butterfly – Karen Oberhauser, US Geological Survey; mountain lion – California

Division of Fish and Wildlife; passenger pigeon – Keith Roper, Flickr; piping plover – Jim Hudgins, USFWS; Puerto Rican parrot – Tom MacKenzie, USFWS; red wolf – John Froschauer, USFWS; Saint Lucia parrot – Christopher Cox; tiger – Davidvraju, Wikimedia Commons; wild turkey – USFWS.

Made in the USA
Monee, IL
04 November 2026

67381846R10154